RESOURCES UNDER REGIMES

NEW HISTORIES OF SCIENCE, TECHNOLOGY, AND MEDICINE

SERIES EDITORS
Margaret C. Jacob, Spencer R. Weart, and Harold J. Cook

PAUL R. JOSEPHSON

RESOURCES UNDER REGIMES

TECHNOLOGY, ENVIRONMENT, AND THE STATE

HARVARD UNIVERSITY PRESS

CAMBRIDGE, MASSACHUSETTS

LONDON, ENGLAND

2004

Library of Congress Cataloging-in-Publication Data
Josephson, Paul R.
Resources under regimes : technology, environment, and the
state / Paul R. Josephson.
p.   cm.—(new histories of science, technology, and medicine)
Includes bibliographical references and index.
ISBN 0-674-01499-5 (alk. paper)
1. Technology and state—History—20th century.   2. Science and state—
History—20th century.   3. Environmental policy—History—20th century.
I. Title   II. Series.
T49.5.J67   2005
304.2'8—dc22   2004051132

CONTENTS

RESOURCES UNDER REGIMES

INTRODUCTION:
NATURE, TECHNOLOGY, AND WORLDVIEW

Earth provides enough to satisfy every man's need but not for every
man's greed.

—Mohandas K. Gandhi

Over the past two centuries the modern state has played a major
role in backing scientific research and development to promote national defense, public health, economic growth, and resource management. Activities that have had a direct impact on the environment include the study of resources through geological surveys; the
funding of expeditions; the support of agricultural and other research; and various flood-control, dam, harbor, roadway, and other
improvement projects. By the early twentieth century, many nations had passed and begun to enforce laws to protect resources
from overuse and to punish polluters. Even before the rise of modern nation-states, kings and queens had established regulations to
protect their lands—forests and the animals living in them—from
encroachment by peasants and poachers.

In this short book, I explore the interrelationship of science,
technology, and the environment in the twentieth century. I focus
on the role of the state in this relationship because modern governments have become major actors in enabling their citizens to gain
access to resources, often through large-scale approaches; create

power generation, transportation, water supply, and other technologies that improve the quality of life; manage pollution problems; deal with the legacy of hazardous waste; and develop legal frameworks for addressing the global environmental crisis at the beginning of the twenty-first century.

A book of this length on a subject of such magnitude will lead some readers to wonder what subjects are missing, but I do not pretend to undertake a complete exposition. The goal, rather, is to encourage exploration of ideas and themes important to environmental history. I offer brief, and I hope provocative, answers to some of the questions I raise, and in other cases I point out where the reader can find his or her own answers to them. In other instances, I merely raise questions to set readers thinking. My intention is to provide a brief, readable essay on the interplay among the modern state, technological systems, and the environment in comparative perspective. I consider the experiences of countries in the industrialized world and the industrializing world. I ask readers to ponder to what extent the nature of the government and economy—authoritarian or democratic, centrally planned or market, colonial or postcolonial—shapes the considerations of political leaders, policy makers, scientists and engineers, and other citizens about how to deal with various environmental and social problems that must be faced in transforming the natural world into the human world.

Environmental history is a relatively new field of research and writing that seeks to understand human interaction with nature over time against the backdrop of the development of human political, social, and cultural institutions and trends. Environmental history is interdisciplinary, bringing together history, geography, anthropology, the natural sciences, and many other disciplines. No one disputes that humans have an impact on the environment. This is, I must say, natural and human of them. But environmental history seeks to place human activities and institutions within nature, not outside some pristine nature that is somehow untouched by human hands.

Where is that pristine nature? Is the Amazon rain forest pristine because few urban dwellers have reached it? What of the indigenous people living within that forest? Are national parks like Yosemite in California that serve hikers and tourists pristine, or the Manuel Antonio National Park in Costa Rica, which was established to protect biodiversity?[1] Environmental history also explores how humans use natural resources for human purposes: harvesting lumber for fuel, building materials, or paper; extracting ore for smelting; or clearing land for farming.[2]

A constant tension exists between those who see a world increasingly on the brink of environmental disaster and those who believe that the risks are overstated. The second group includes scientists, policy makers, and businesspeople who express the reasonable desire to harness resources for the benefit of the world's citizens and claim that resources are extensive enough, if managed scientifically, to meet the needs of present and future generations. Many among this group insist that the rights of individual property owners ought to take precedence over the needs of society as enforced by the state—for example, through eminent domain. These theorists point to the cases of the former Soviet Union and China as evidence of how nature can be destroyed when the state plays too great a role in control and distribution of natural resources. Such observers often share the belief that regulation can result in needless interference and may be based on poor science. They believe that market mechanisms, more than any other means, will seamlessly set scarcity values (that is, prices for goods and services that the market determines through supply and demand) and prohibit unbridled use of resources. Still others, chillingly, discount the existence of global warming and call for still more research to prove it.

Other observers see little need to debate the evidence any longer.

Consumption patterns, especially among the world's democracies, have rapidly depleted resources, many of the world's forests have been burned or felled, water and air pollution threaten the globe, and environmental disasters have destroyed flora and fauna, as the increase in extinct species reveals. For centuries people have dumped offal, tanning chemicals, and lead and other heavy metals into the ground, into rivers and lakes, and until recently into public water and sewage systems. In the twentieth century, with the rise of the modern chemical industry, the production of pesticides, petrochemicals, polychlorinated biphenols (PCBs), and other chlorinated hydrocarbons has grown astronomically, as has the problem of their proper use, handling, and disposal. The developers of many of these chemicals initially touted them as wonder workers, for example, PCBs as insulators and coolants for electrical transformers in the burgeoning power industry in the 1920s and 1930s. The dangers and costs of these wondrous chemical compounds were discovered only much later and have required strong authority to investigate their impact and force their cleanup. Again, PCBs are a good example: after lengthy legal proceedings, the U.S. Environmental Protection Agency (EPA) required the General Electric Company to remove PCBs from the Hudson River.

If in the member nations of the Europe-based Organization for Economic Cooperation and Development (OECD) more than 70 percent of hazardous waste is disposed of in landfill sites, about 8 percent is incinerated, and 10 percent is recycled, what happens to the rest? Thousands of tons and millions of liters of low- and high-level radioactive waste continue to be stored temporarily—and inadequately—all over the world, including in unstable regions. The situation is particularly serious in the former Soviet Union, the rest of Eastern Europe, China, and the numerous poor countries of Asia and Africa. In 1991 Poland alone produced 128 million tons of industrial waste, nearly all of it from the mining and metallurgical industries.[3] Entire regions of Russia have become industrial deserts where nothing, not even the hardiest grasses, will grow.

Many environmental disasters are so severe, so horrifying, so destructive to nature and humankind that they have entered our daily lexicon. They include the Bhopal chemical disaster in India, which killed more than six thousand people; the Chernobyl nuclear plant explosion in April 1986; the *Exxon Valdez* oil spill; the ground and water contamination in Love Canal, New York, where the discovery of dangerous chemicals necessitated the emergency relocation of town residents; the Minimata Bay disaster in Japan, where the illegal dumping of mercury destroyed the lives of an entire fishing community and killed more than fourteen hundred people; and the ongoing burning of the world's last rain forests. At the root of each of these disasters is human manipulation of natural processes through science, technology, and industry.

Environmental problems that lack specific names are no less crucial to redress: rapid deforestation; logging on public lands, in the name of preventing fires or creating jobs; the increasingly rapid disappearance of species and loss of habitats; the inadequacy of fresh water for most of the world's population; disease vectors that follow those of pollution; greenhouse gas production and global warming; and holes in the ozone layer. Can laws alone address these issues? Are international treaties required? Must we change our patterns of consumption and live more frugally to avert environmental disaster? While the causes of some of the problems of resource use, pollution, and waste disposal remain in dispute, there can be no doubt that they require our attention and remediation, and that open governments and well-informed citizens are most likely to select the best means to tackle the problems without requiring a return to an early industrial lifestyle.

Paradoxically, the roots of many of these environmental problems lie in the very effort to place management of resources on a scientific basis that was intended to produce improvements in the quality of life. Often lost in the discussion of costs, benefits, laws, waste, and pollution is the role that science and technology have played in giving us the tools to manage resources scientifically and simultaneously leading us to believe that those resources are inexhaustible. When we move ahead too quickly, we assume we will find solutions to the problems we have created through science and technology. Over the past four centuries scientists have developed an ever more complex understanding of "nature" on many levels, of plants and animals, rivers, streams and lakes, tides, and weather and climate and their impact on ecosystems. Researchers have gathered data on rainfall and temperature, on river flow, and on plant growth over scores of years and can now discuss patterns with relative certainty. On the basis of extensive research, scientists have attempted to make nature seem more orderly and predictable. They have drained fens, reclaimed land, and established industrial forests.

Girded by the growing hubris that we fully understand biological processes and possess unbounded engineering skills, we have turned rivers known for their seasonal flow into hydrological machines that store water in reservoirs according to estimates of annual flow and release it through hydroelectric power stations, canals, and irrigation systems when we order it for purposes of agriculture and power generation. With hybridization, pesticides, and herbicides we have created monocultures of high-yield cash crops. The systematic effort to improve on nature that began with the scientific revolution has culminated in the creation of new plants and animals through genetic engineering. Because of its now regular, predictable productivity, in agriculture, silviculture (industrial forestry), and aquaculture (industrial fisheries), we produce adequate food to meet the world's needs, and we consider ourselves capable of mastering nature itself. And given the quality of life Europeans and North Americans have achieved and the bounty of production, why should we think otherwise?

HOW DO DIFFERENT SOCIETIES VIEW THE WORLD AROUND THEM?

The first step toward exploring these issues is to understand the impact of worldview on the changing interrelationship of technology, the environment, and the state. By "worldview" I mean how we believe the world around us is structured and how it functions in both physical and symbolic ways. For example, the transition from the geocentric (that is earth-centered) view of the universe to a heliocentric view (which locates the sun in the center and the earth and other planets orbiting it) involved far more than a change in the place of the earth in the universe. It entailed an entirely different view of the place of men and women in that universe. Before, the universe was small and finite, and only through religious practice might human beings come to understand its workings; this comprehension was important more for the afterlife than for day-to-day life here on earth. When the universe was understood to be vast, perhaps infinite, humans and the earth were no longer at the center of God's creation. It became the scholar's role to study the universe, gain understanding about the mysteries of its workings, and apply that understanding to improve upon those workings, and ultimately to control the earth's mines, tame its forests, and change the course of its rivers. That is, there arose a new view of man and woman's position in the universe vis-à-vis God and nature, with humans gaining preeminence over what some scientists called God's sensorium, to change it, as they believed, for the better.

The transition from a heliocentric to a geocentric universe occurred during the Scientific Revolution (1500–1700). Granted, large-scale and state-supported transformation of nature predated the Scientific Revolution. The Romans had built an extensive network of aqueducts to supply their empire with water. The lead pipes they employed slowly poisoned many Romans. (Lead pipes were outlawed in most countries only in the second half of the twentieth century.) Mining, artillery, and other technologies also date to the Roman period. Likewise, pond aquaculture, mills, irrigation systems, and the systematic exploitation of forests all predated the scientific revolution by hundreds of years. But during the Scientific Revolution the effort to control and adapt nature to human designs and desires took on a more systematic and eventually grandiose character. In Mesopotamia and Egypt, whose civilizations were dependent on rivers, the Euphrates, Tigris, and Nile, the inhabitants had focused their efforts on the control of floods, on the building of dikes and irrigation systems. The draining of the fens in England, the reclamation of land from the ocean in the Netherlands, the establishment of state-managed forests in Russia, Germany, and elsewhere, the creation of pumps that enabled miners to go deeper into the earth than before—all these achievements marked the culmination of millennia-long experience of modifying nature.[4]

Three fundamental changes characterized human designs on nature from the sixteenth century onward. The first involved the increasingly science-based approach to resource management. For instance, by the end of the eighteenth century, foresters had begun systematic investigation of various techniques to improve lumbering productivity, and by the end of the nineteenth century, universities had established experimental fields for corn and other crops. The second concerned the growing role of the state in the inventory and control of natural resources, and in the promotion of transformative projects. The third was the rise of the so-called mechanistic worldview in connection with a new conception of the notion of progress.

During the Scientific Revolution, the discoveries of such scientists as Copernicus, Galileo, Descartes, and Newton led many people to believe that a mechanistic, or machinelike, model could describe the universe. According to those who embraced the mechanistic worldview, most, if not all, of the phenomena that occur in the universe, including biological and chemical phenomena, can be explained through laws of Newtonian physics. The rotation of the planets, the movement of stars in distant galaxies, the trajectories of bodies on earth, the tides, and weather are all governed by those laws. The simplest statement of this worldview is that the universe is a machine or clock that God created, wound up, set in motion, and left for humans to use.

Over the next two centuries scientists explored the extent to which mechanical laws applied to life phenomena as well. To a certain extent, physico-chemical explanations took root in the biological sciences. Some of the most extreme of their adherents—the so-called reductionists and mechanical materialists, for example—believed that life itself could be explained by physics. As Jacques Loeb, an American embryologist, wrote in 1916, "We eat, drink, and reproduce not because mankind has reached an agreement that this is desirable, but because, machine-like, we are compelled to do so." Loeb even included ethics in his view of life: "Not only is the mechanistic conception of life compatible with ethics: it seems the only conception of life which can lead to an understanding of the source of ethics."[5] Carolyn Merchant offers the provocative suggestion that the mechanistic worldview embodied clear notions of manipulating nature as never before; nature had become something to be "controlled and dissected," something machinelike and also to be controlled by machines. The result, she writes, has been "the death of nature."[6] During the scientific revolution such scholars as Sir Francis Bacon urged us to study nature because knowledge is power, but they never intended that we should destroy nature.

Although many scientists took a more balanced view of the extent to which physico-chemical laws could provide answers to questions of biology, environment, and resources, others, including rulers, policy makers, and businesspeople, continued to act as if nature could be controlled mechanically, its rivers and lakes altered by dams and locks or dredged to operate in a predictable, orderly fashion; its forests planted in neat rows as required by the military or the housing and paper industries; its wetlands drained and the reclaimed land made "more perfect" than God had intended, in that it had been put to productive use.

Not only had nature become more machinelike, but it was now society's role to improve upon it. This new view of "progress" gained currency in Europe and North America during the Enlightenment, in the second half of the eighteenth century. The Enlightenment was a conscious proclamation, based on notions of equality, the worth of the individual (and, among many thinkers, the advantages of democracy over monarchy), and the power of science, of the nearly limitless capacity of reason to improve the conditions of humanity. The importance of the Enlightenment for the present discussion is that the notion of progress meant that humans no longer needed to wait passively for the seasons to change or for rivers to flow or forests to grow; rather, human beings were enjoined to apply science and engineering to obtain regular, orderly production from nature. Further, the fruits of this process would benefit all of humanity. Ultimately, that production, including in agriculture, fisheries, and forestry, acquired an industrial basis.

The French *philosophe* Marie-Jean-Antoine-Nicholas Caritat, marquis de Condorcet (1743–1794) best expressed this relation of progress, society, and the natural world. Condorcet believed in the imminent and almost limitless progress of human reason and the sciences. Human history had progressed from the civilization of primitive hunters and gatherers to agricultural and now industrial societies. The "mechanical arts," or what we today call industry, had "no other limit than the reach of the scientific theories on which they depend . . . Instruments, machines, looms will increasingly supplement the strength and skill of men; will augment at the same time the perfection and the precision of manufactures by lessening both the time and the labor needed to produce them. Then the obstacles that still impede this progress will disappear, and along with them accidents that will become preventable and unhealthy conditions in general, whether owing to work, or habits, or climate." Thanks to human progress, "a smaller and smaller area of land will be able to produce commodities of greater use or higher value." Condorcet wrote: "Thus not only will the same amount of land be able to feed more people; but each of them, with less labor, will be employed more productively and will be able to satisfy his needs better." With the pursuit (and aid) of science, and the march of progress, nowhere more than in the Europe of his day, had come the revolution of the Enlightenment. The nations of Europe, the most progressive, would bring civilization to the vast lands inhabited by "savages." Condorcet added, "The progress of these peoples is likely to be more rapid and certain than our own because they can receive from us everything that we have had to find out for ourselves, and in order to understand those simple truths and infallible methods which we have acquired only after long error, all that they need to do is to follow the expositions and proofs that appear in our speeches and writings." He concluded, "The time will therefore come when the sun will shine only on free men who know no other master but their reason."[7] Today we might mistake Condorcet's views for overweening confidence but for his time he revealed a faith that human beings could solve the major problems facing their societies. Perhaps it required the conjunction of hubris and state support of the engineering profession for humans to embark on the large-scale projects of the twentieth century to transform nature. And there is no question that, at least to some degree, Condorcet's visions of a rational modern society producing a seemingly unlimited variety of important commodities with great benefit to citizens and leaders alike has been achieved.

A crucial factor in the ability to manipulate nature was the growing role of the modern state. Whether monarchies, democracies, or authoritarian regimes, states created various departments or ministries to take stock of resources and calculate their availability. In the early 1700s Peter the Great established imperial control over Russia's forests to secure lumber for his navy. Over the next century Catherine the Great and other tsars created a forestry department in the government to make an inventory and train specialists to attend to the forests. One of the first acts of the U.S. government was to establish the U.S. Geological Survey. The U.S. Army Academy at West Point, New York, trained a corps of engineers, the elite graduates of West Point, who to this day are among the most significant canal builders, earth movers, shore rebuilders, river alterers and harbor dredgers in the world. Under Kaiser Wilhelm in the early twentieth century, Germany established the Kaiser Wilhelm Gesellschaft, a series of institutes in major fields of science and engineering, to secure the lead in fields important to the nation: economic strength (especially in the chemical industries), military security, and public health. In the late nineteenth century the Norwegian parliament created a fisheries inspectorate that supported specialists and oceanographers gathering data on ocean temperature, chemistry, and currents, on fish populations, migration patterns, and life cycles; this information benefited coastal fishing communities. The Soviet Union created literally hundreds of research institutes to build up the economy and the military.

Through financial support, direct organization, establishment of ministries or departments, creation of public health services, and so on, the state has played a central role in issues of importance to the environment. This role was enhanced by the belief that scientific and engineering expertise would facilitate control over nature. By the late nineteenth century, during the Progressive Era in the United States, scientists and engineers had come to believe that using the objective tools of science, they could provide simple answers to complex questions about how to manage such finite natural resources as water and allocate them to industry, agriculture, recreation, and other competing interests, for present and future generations. A partnership between government and science based on this kind of thinking developed in different nations—in different ways—in the twentieth century. There is no question that science and technology have helped modern states make significant improvements in public health (by lowering mortality rates and raising life expectancy), environmental safety, and economic performance (by providing more jobs, and safer, fresher food and better diets, for example). A question for us to consider is whether science and technology have benefited powerful states, businesses, and members of society equitably.

Joining the modern state in promoting science and technology were industry and higher education. Capitalist owners of factories contributed strongly to the rise of the modern laboratory, both in the industrial enterprise and at the university, in pursuit not only of profit (although largely so) but also of improvement in the quality of life and of knowledge for the sake of knowledge. Some of these enterprises became major players in issues that are important to environmental history—Dupont and Dow in chemical biocides, Monsanto in genetically engineered crops and hormones, Westinghouse in nuclear reactors, General Electric in PCBs.[8] As for the university, the best examples are land grant universities and colleges such as University of California, Davis, and Iowa State University. On the basis of research conducted at those universities, and often in corporations connected with their research programs, the quality of food has improved significantly, the bounty of the harvest has grown, and food is cheaper in real prices than a few years ago. University systems in England, Sweden, Germany, and elsewhere have been closely involved in research on the management and exploitation of natural resources. In this book we consider the environmental benefits and costs to society of academic and industrial research—for example, as regards the extensive use of chemical pesticides, herbicides, and fertilizers; increased soil erosion; and hundreds of millions of dollars' worth of research on modern agribusiness practices, feeds, seeds, and techniques, including the mechanization of harvest of apples, asparagus, onions, tomatoes, and lettuce. Does this research benefit the consumer, the small farmer, or the agribusiness?[9]

TECHNOLOGY AND ENVIRONMENTAL DEGRADATION IN THE MODERN WORLD

Beyond those who welcomed and praised the potential for a partnership of science, engineering, and the state to transform nature, many groups have lamented the place of technology in its transformation, either directly or indirectly. In early nineteenth-century England the Luddites went about smashing looms and other equipment in mills, lest the worker be displaced in the factory by machines. Since that time, people who question the pace of assimilation of technology in society or wish to turn back the clock to a simpler age have been referred to as Luddites. But critics of technology are usually more sophisticated than the Luddites. The two major groups of critics are those who worry about the uneven social and environmental costs of development (including contemporary critics of "globalization") and those individuals who advance the notion of appropriate technology.

Many analysts see the roots of extensive environmental deterioration in the inappropriate application of modern Western science across societies and ecosystems, especially in Africa and Asia. For example, adherents of dependency theory base their criticism of development on the Anglo-European paradigm of industrialization and of the employment of high technology to bring about rapid modernization in countries that wish to industrialize. The critics rightly point out that the use of Western technology and aid links traditional societies to modern ones in ways that are often damaging to existing social and cultural arrangements. The introduction of high-yield crops, for example, requires dependent nations to take out loans to pay for the seed and the fertilizers and pesticides required. National and local officials must require peasants to abandon traditional forms of cultivation in the hope that farmers will produce cash crops to earn export income to pay off the loans. Peasants gain little knowledge about the new crops, traditional indigenous agriculture falters because investment and research to develop it are lacking, and the peasants become marginalized.

Supporters of development theory, by contrast, assume that the Western model, based on high technology and modern scientific research, will transform a traditional economy into a modern, capital-intensive one—with new forms of employment and large export markets—more efficiently than any other approach. They believe there is one best path toward economic "maturity," even though this path often turns out to be ethnocentric, benefiting only some groups at the expense of most others, and urbanist, benefiting city dwellers at the expense of rural residents. Proponents of development are convinced that a top-down approach will work, even when it ignores local interests and cultures. Yet this path toward development is technocratic, requiring reliance on specialists from abroad or at least on a foreign-trained technical elite, both of whom place great store by modern science and little store by local knowledge.

Crucial questions that we shall consider about development, technology, and the environment are, What is development? Who benefits from it? Who tries to control it and for whom? What was the impact of development, fostered by colonialism, on environment and society? In a study about modern high-yield aquaculture in the Philippines, Philip Kelley writes, "Development signifies the processes in which existing systems of production and reproduction in society give way to new ones . . . A development strategy always implies a shift in power relations based on class, gender, access to environmental resources and other axes of inequality." He urges us to consider the impact of development not only on some macro level of income production but "at the scale of the household, family, individual and village where the implicit biases in development strategies are translated into the harsh reality of poverty and oppression."[10]

In a similar way, critics of globalization point out its benefits primarily for the wealthy nations whose scientific and technological know-how, financial wherewithal, and political clout enable them to control resources and production processes around the globe. The critics argue that globalization entails the export of environmental problems to the third world. This occurs through the transfer of dangerous technologies and processes abroad, including hazardous-waste disposal. The critics refer to the abuse of human rights and workers' rights through globalization. This abuse occurs through the export of the most unpleasant and lowest-paying jobs abroad to countries where concerns about worker safety are poorly developed and unions have little power. And globalization exacerbates inequality and poverty between the haves and the have-nots.

A crucial issue, according to the critics, is that the major proponents of globalization have the greatest power and resources to act. Roughly half of the largest economies in the world are corporations, and half are nations. The sales of the corporations far exceed the incomes of the billions of people living in abject poverty. Indeed, the income of the poorest inhabitants of the globe has declined, while that of the wealthiest has grown—all under the banner of globalization.

Finally, there is the school of appropriate technology. Supporters of appropriate technology (AT) argue that technology is not inherently dangerous in its transformation of social, economic, and environmental systems. Rather, many technologies, particularly if they are small in scale, not capital-intensive, and perhaps even based on indigenous knowledge and experience, provide a rich foundation on which to build economically and environmentally sound societies. Promoters of AT also offer an alternative view of "progress," one that stresses the importance of maintaining traditional social structures, in part by adopting small-scale modifications of existing agricultural and manufacturing arrangements.

For example, rather than focus on high-yield cash crops, the farmer should gain local institutional and financial support to develop food crops. And rather than rely on the forests for charcoal and firewood, the farmer might turn to biomass. A good example of a technology that uses biomass (dung, urine, and other biological waste) is the biogas digester that produces methane for cooking, illumination, and power, such as the one developed jointly by German and Kenyan specialists.[11] Biogas digesters are also widely used in China. Supporters of appropriate technology completely reject such large-scale projects as hydroelectric power stations that require the uprooting—not to mention the inundation—of entire communities, when biogas could supply local needs.

Appropriate technology took its inspiration from Ernst Schumacher's *Small Is Beautiful* (1963). According to one author, the AT method "attempts to recognise the potential of a particular community and tries to help it to develop in a gradual way. This development is based on local resources and progressively builds up the skills of the community. It is essentially rural-based rather than urban-based as is the Western technology." Appropriate technology focuses not on extraction of resources or transformation of nature but on improvement in the quality of life of the people, maximal use of renewable resources, and creation of local jobs, using local skills, resources, and finances that are compatible with local culture and local wishes and needs.[12]

An important aspect of appropriate technology is the belief that local, indigenous knowledge offers economic and social continuity that provides stability when different techniques are introduced in agricultural and other traditional communities. To take one example, Vandana Shiva explains that to be civilized in India means to practice Dharma; *Dharma* means "the stabilizer." What is the source of stability in human societies? Shiva argues that the stability of civilized societies is related to the stability of natural resources and ecology. "Dharma consists in restricting use of resources to satisfaction of basic needs because using resources beyond one's needs would be appropriating the resources of others." Justice, ecology, and social stability are thereby linked, for restraint in use of resources is a precondition for peace and justice.[13]

Mohandas K. Gandhi worried about the unsettling impact of Western technology on traditional Indian society. He said, "Politics is the handmaid of commerce. Indian history provides an apt illustration of it. In the heyday of our cotton manufactures, we used to grow all the cotton for our needs. The cotton seed was fed to the cattle which provided the health-giving milk to the people. Agriculture flourished." But because of the establishment of British monopolies on Indian textiles and the industry the British imported, "all that is gone now. India is today naked." He believed that engines and machines, which put thousands and thousands of hands out of work, were used merely out of greed, not to supply water or alleviate poverty.[14] Gandhi wrote that India had opted for another form of development not because of "technological inadequacy, but because of ecological sophistication." He observed, "We have managed with the same kind of plow as existed thousands of years ago . . . It was not that we did not know how to invent machinery."[15] Thus, while many observers point to significant improvements in the quality of life, great achievements in agriculture and public health, and the power of modern science behind these developments, such thinkers as Gandhi criticize the Western development paradigm. They believe that the industrializing world which grew out of the scientific revolution and the Enlightenment has had disproportionately large social and environmental costs.

Gandhi's views are widespread among persons who favor alternative development strategies that appear to be more consonant with existing social and ecological structures. Konrad Schliephake, a specialist in agricultural economics, writes, "Many times, as a researcher and consultant, I have asked my local partners to listen more to the old men, the old peasants who have possessed their knowledge for generations and who know the details of their soil, their plants, and their climate." Irrigation changes things, often for the better. But Schliephake worries, "However, we should not forget the indigenous traditions."[16] At the same time, we should not idealize the small-scale or short-lived impact on the environment of indigenous people or those without modern technology. All cultures interact with nature, turning to slash-and-burn agriculture or fire to clear forest. There has been no pristine nature since humans appeared.[17] And finally, we should not idealize local knowledge as better than science merely because it is local. Many people mix magic and religion, creating belief systems ill equipped to deal with the challenges of medical care, economic change, or environmental degradation.

Some writers mistakenly believe that Karl Marx and Friedrich Engels criticized the power of technology to transform nature. True, they objected to the way capitalists employed technology to exploit and alienate workers. But theirs was a doctrine of progress, according to which technology would liberate the worker from thankless toil and onerous menial labor and enable the creation of a utopian society. They clearly envisaged in communism an advanced industrial society in which workers were truly free. The modern factory would be a glorious, clean, well-illuminated facility powered by electricity. Machines would do all the difficult and heavy labor and help produce the basic necessities, while people could choose a lifestyle based on their interests. Marx and Engels did not directly address the relation of science, engineering, the natural environment, and the state in any single work. But they and their followers recognized the powerful effect of the development of the productive forces—machines, tools, factories, the "means of production," and laborers themselves—on the ability of humans to control nature. Hence, technology was the source of alienation in capitalist society, but the key to liberation in socialist society. We shall consider the People's Republic of China and the former USSR, two archetypal Marxist regimes that also attempted to embrace modern technology to transform the natural environment.

TECHNOLOGY, THE ENVIRONMENT, AND THE STATE

How does the particular form of the state and its relation with businesspeople, specialists, and other citizens aggravate or ameliorate environmental problems? I discuss development patterns in three different kinds of states—democratic or pluralist, authoritarian, and postcolonial. I argue that pluralist states have developed institutions that are far more responsive to environmental problems than have the other two types of regimes, largely because of broader access by citizens to information about the appropriate path of development, and because of the creation of legal, scientific, and other institutions to mitigate environmental problems. It is true that democratic regimes—mainly the countries of North America and Europe—pushed by various economic and political pressures and pulled by consumer demand—often attempt to export the costs of resource use and waste disposal to postcolonial regimes.

I ask readers to confront the paradox that authoritarian regimes, although claiming to exploit resources more efficiently and equitably for their citizens and asserting that only their institutions can protect the true interests of the citizens and nature against the unbridled forces of the market and the search for profit, have environmentally and socially costly development patterns. No less than pluralist regimes, and in fact more so, authoritarian governments serve the interests of privileged party members and their allies. Marxist regimes, for example, claim to represent the worker, even as the worker toils in dreadful conditions and rarely reaps the benefits of state ownership of the means of production available to Communist Party members.

Last, I explore the thorny issues of why countries that have gained independence after lengthy colonial experience continue to experience challenging environmental problems that range from polluted water and soil erosion to rural outmigration, crushing imbalances between the industrial and agricultural sectors, and the pandemics of AIDS, tuberculosis, and other diseases. What contributes to the difficulties postcolonial nations have in addressing the serious environmental problems that affect their citizens? Are there technological solutions to these problems?

The reader will observe several constants in my arguments. First, I argue that technology is inherently political, that tied into decisions about which technologies to use to develop resources or to deal with environmental problems or to improve the standard of living are political, economic, and social desiderata. The decision to build a hydroelectric power station is based on more than a calculation of annual river flow, the areas geologically most conducive to building dams, and potential electricity demand—that is, on more than hydrological, geological, and economic calculations. Political (legislative, regulatory, and judicial) and economic (banking, manufacturing, and construction) institutions must be involved in the decision to build a hydroelectric plant. Social questions about how to relocate people whose homes and property will be inundated by the waters that will back up behind the dam, what kind of compensation those citizens should receive, where they will live, and what will happen to the history of their community (schools, cemeteries, and so) must also be addressed. Questions of equity—the distribution of the costs of state-sponsored technological change among all members of society—must be addressed.

Second, I argue that large-scale development projects will usually be more costly and have more irreversible impacts than small-scale development projects, and that large-scale projects run a greater risk of overwhelming democratic institutions and concerns than do small-scale, locally based projects. To the extent that local governments and groups have had the ability to question, if not overturn, decisions made in government bureaucracies or engineering firms that are distant from them temporally and psychologically, it was less likely that environmentally damaging programs would be approved or regulations ignored or overturned. For example, concerning the Clean Air Act in the United States, the federal courts have usually sided with environmental groups and a growing number of state governments in the Northeast to prevent fossil fuel–based power-generating companies from avoiding cleaner-burning technologies. The courts took this position even though the administration of George W. Bush supported the companies in their efforts to avoid buying available modern equipment and abandoned legal proceedings against a number of power producers who had violated existing laws.

Of course, regulatory powers and processes differ from country to country, but readers will see that regulations in some countries might be employed with great effect in other nations also to address environmental issues. That is to say, the state can make a difference in confronting the environmental problems connected with technological excess in democratic regimes. By definition, pluralist regimes are more responsive to input from citizens. They have a well-developed civic culture that makes it easier for political institutions to arrive at reasonable environmental (and other) policies.

Third, the evidence indicates that both authoritarian regimes and postcolonial regimes are ill equipped to deal with the problems of sustainable development, preservation, conservation, mitigation of environmental problems, and regulation of manufacturing processes, construction, and waste disposal. For authoritarian regimes, even if the state appears capable of decisive intervention to protect the environment, the raison d'être of the government is the promotion of rapid economic development in the interests of state power, military capability, and international stature. While the state assumes that it has a place in history for the long run (witness the self-proclaimed Third, or "thousand-year," Reich), its development programs stress short-term economic achievements (immediate industrial production or military power), not long-term social and cultural ones (the preservation of nature for future citizens). Similarly, the citizen is nearly powerless, whether acting individually or as part of a group, to protest against state decisions that may have a disastrous environmental impact. Even the input of experts (scientists, engineers, and economic planners) in decision making is circumscribed by state power. In many cases, the expert has been co-opted by the state. Any balance that exists in democratic regimes between the private and public sectors, which permits careful weighing of individual rights versus the social good, is absent in authoritarian regimes. The authorities in those regimes assume that *they* should determine what balance, if any, is achieved.

As for postcolonial regimes, the reasons for their persistent environmental problems—and the absence of technological solutions to them—range from the historical to the political and economic. The legacy of exploitation of resources for the benefit of the colonial power, the forced replacement of traditional forms of subsistence agriculture by cash crops, and the destruction of existing political institutions have contributed significantly to the extensive problems facing postcolonial governments, especially in Africa and South Asia. Yet the political authorities in many of those countries seem to have no interest in changing the patterns of resource exploitation. The political and ethnic elites often pursue development of markets for new products to benefit themselves, at the expense of peasants, farmers, and the environment within their borders. International banking organizations, multinational corporations, and governments that offer foreign aid have contributed to the persistent problems by promoting technological solutions (such as "green revolution" high-yield crops and chemical- and water-intensive agricultural techniques) that are usually accompanied by soil erosion, improper use of biocides, and destruction of traditional social structures. Urbanization, industrialization, and modernization of agriculture have not been the panacea that their supporters promised. For example, when peasant families are broken apart by the destruction of traditional agricultural institutions, men migrate to the cities in search of jobs, meanwhile leaving women and children behind to take care of the farms. In the urban centers, the men enter the dangerous territory of modern disease vectors. In other words, AIDS, follows their migration patterns. Lacking national engineering and scientific traditions, and having weak civic culture, postcolonial regimes are often unable to explore any alternative to development or to address the environmental problems they face with the necessary vigor and depth of analysis.

I once believed that a petrochemical plant, a nuclear power station, and a highway were essentially the same anywhere in the world, irrespective of the country in which they were built. The laws of physics, geology, chemistry, and biology, the engineering considerations—strength of materials, the weighing of inputs and outputs, and so on—are the same across borders and are based on the universal laws of science. But questions of science and engineering—questions of what *is* and how it operates—are not questions of what *ought to be* and how to get there. Where the state allows a full discussion both of what is and of what ought to be, where the state openly considers the balance between the costs of regulation to business, the costs of unfettered development, and the costs of postponing difficult decisions about resource use, there is less likely to be extensive, irreversible environmental devastation. In open systems, the technologies and approaches employed to manage and develop resources are less likely to contribute to social and environmental disruption. By contrast, where the state is too powerful and seeks to promote its power, or where the state is too weak or uninterested in promoting balanced use of resources, the costs to all citizens are greater.

Another goal of this book is to understand the sources of environmental deterioration that plague all the world's inhabitants in all political systems in the twenty-first century, and to encourage readers to explore solutions to the problems. The citizens of the wealthier, industrialized countries seem to avoid many of these ills. They can export their wastes and their dangerous jobs, at the same time embracing a highly consumptive, some would say wasteful, lifestyle. Having turned in the direction of service-based economies, these countries can import industrial goods, the production of which can be highly polluting. Industrialized nations can afford more energy-efficient, "greener" technologies that harvest, process, and distribute goods and services with less waste than older technologies do—and can afford regulations to promote green technology.

The citizens of wealthy nations not only use far more resources per capita than do the citizens of poor nations, but they pollute far more than developing nations. Both pollution and the rapid depletion of resources contribute significantly to the destruction of nature. But developing nations will use more and more fossil fuels in the attempt to industrialize rapidly. China and India are likely to become the worst offenders. Still, the leading polluter and user of resources remains the United States which, with a twenty-fifth of the world's population, consumes a fifth of the world's resources and produces nearly a quarter of its greenhouse gases (and two-fifths of all emissions by industrialized countries).

The industrialized nations can also afford stricter environmental protection laws. Many analysts and policy makers have such a great distaste for any government regulation that they believe that the market should somehow magically determine the extent of regulation and fines to control pollution. These critics argue that government regulation indeed prevents or slows economic growth. The fact that the world's largest and strongest economies are those with the most vigorous environmental protection laws calls this assertion into question. That the vast majority of citizens in those countries also prefer environmental protection laws to pollution or vague appeals to freedom of choice should also inform the discussion whether to relax or strengthen laws. And the leaders of the poverty-stricken countries of the world—for example, Sudan in North Africa, Bangladesh in South Asia, and dozens of others—would welcome the opportunity to have laws as strong as those in Sweden, Germany, or the United States, and ecosystems as clean.

In Chapter 1 I provide an overview of the interconnection between technology, environment, and the state in democratic regimes. In the twentieth century the state played an important role in environmental issues by financing projects, regulating technologies, and establishing pollution control laws. The ideology of political and business leaders and engineers often overlapped, as they pursued policies to promote the development and extraction of natural resources and the transformation of the land and water. In part because of the growing insistence of the public, governments turned to various environmental protection laws and resource management regulations to protect those resources for present and future generations.

In Chapter 2 I turn to authoritarian regimes—generally speaking, those which have a single political party, limited access by individuals to the policy process, and coercive means of dealing with dissent, including environmental activism. In the economies of these nations, which tend to be centrally planned, pollution levels are often extremely high, and the absence of regulation prejudices industrial safety and environmental protection. Unlike pluralist regimes, authoritarian regimes—whether in Stalinist Russia, Nazi Germany, or Brazil under a military dictatorship—link ideologies of progress with a group or people whose interests, the leaders determine, have heretofore been denied—for example, workers, under Marxist regimes. The pursuit of "progress" in these nations has exacted a great environmental and social toll.

In Chapter 3 I turn to discussion of southern-tier nations, primarily those of Africa and Asia, whose economies have focused largely on agriculture, and the impact colonialism and industrialization have had on them. Does the blame for poverty and environmental degradation in developing nations lie with colonialism in all its manifestations, including the use of Western science and technology to promote modern agriculture and industry, or with outdated agricultural and forestry practices, unabated population pressures, and other indigenous factors?

In Chapter 4, the conclusion, I discuss the changing range of environmental issues confronting governments and their scientists in the post–cold war world and examine some of the major international treaties and conventions that are intended to address the serious environmental that the world's citizens face—for example, global warming and ozone depletion. I explore the notions of sustainability and biodiversity as definable and reachable goals. Finally, I consider whether any "appropriate" technologies offer an alternative to the resource-intensive and environmentally costly technologies that are the mainstay of most nations. Throughout, readers should seek to understand the ways in which worldview, state power, technology, and the environment are interrelated.

CHAPTER ONE

THE MODERN STATE, INDUSTRY,
AND THE TRANSFORMATION OF NATURE

> To have risked so much in our efforts to mold nature to our satisfaction
> and yet to have failed in achieving our goal would indeed be the final
> irony . . . The truth, seldom mentioned but there for anyone to see, is
> that nature is not so easily molded.
>
> —Rachel Carson, *Silent Spring*

In 1661 John Evelyn, Commissioner for Improvement of the
Streets, wrote, in describing London, of "that pernicious Smoake
which sullyes all her Glory, superinducing a sooty crust or furr
upon all that it alights, and corroding the very Iron-bars and hard-
est stones with those piercing and acrimonious Spirits which accompany its Sulphure."[1] Over the next three centuries that "pernicious smoke" increased in quantity and toxicity. Dangerous wastes
from industries and municipalities were dumped haphazardly into
waterways or onto the land, or at most carted elsewhere for burial
close to the surface. In those centuries, as the dangers to modern industrial life from pollution grew clear to all but the most short-
sighted observers, scientists and engineers became ever more skilled
at tapping natural resources. Using large-scale technologies, factories, and power plants, manufacturers set out to produce goods
on a scale previously unimaginable. Workers mined ores deep in
the earth at great risk to themselves, in the process disfiguring the

landscape with machines and explosives and leaving behind great scars and polluted waterways. Those workers clear-cut the forests and dammed rivers willy-nilly, rarely encountering opposition of any sort except from those affected downstream. The result has been the large-scale transformation of nature and a legacy of hazardous waste that at the end of the twentieth century seemed intractable.

In pluralist regimes, scientists and engineers, in concert with politicians and ordinary citizens, grew increasingly adept at developing the tools to manage the legacy of industrialization, urbanization, and modernization. These constituencies recognized that the negative effects of resource exploitation and of the production process could be mitigated in a variety of ways. These included more efficient methods of production. If lumber mills of the nineteenth century discarded as waste 30 to 40 percent of the tree, mills of the late twentieth century used virtually all of it in products ranging from plywood to particleboard. Bark, scraps, chips, and dust now fuel boilers producing clean heat and electricity.

Another step in improvement was the creation of rules, procedures, and regulatory agencies through which the state, acting in the interest of present and future generations, established fair market values for the use of resources, insisted on proper handling and disposal of wastes, made violation of regulations punishable by fines and imprisonment, and provided access to decision making and legal considerations open to all citizens—businesspeople and regulators alike. Pluralist governments (those with multiparty systems, universal suffrage, and a well-established legal framework to ensure due process) generally encourage public involvement in decisions about resource management and environmental protection. Reluctantly at first, then increasingly, in response to growing public concerns—especially about air pollution (in the 1940s), fallout from weapons tests (in the 1950s), and pesticides (in the 1960s)—elected representatives in Europe and North America passed legislation to protect the environment and to ensure public access to the policy process for environmental protection.

Pluralist governments more successfully than other regimes met the challenges of regulating production and consumption, without adopting the heavy-handedness of authoritarian regimes or threatening the sanctity of private property or civic rights. Citizens' involvement, through the agency of nongovernment organizations (NGOs) or by individuals, came to be founded on the belief that promoters of technology bore the responsibility for demonstrating the safety and efficacy of technology before its introduction. No longer must citizens prove a technology unsafe well after its introduction. The technology assessment process became more open, and data demonstrating safety had to be made available. Of course, there must also be corporate responsibility, in that firms promoting new processes or techniques which have an environmental impact must honestly and openly assess that impact.

Already during the eighteenth and nineteenth centuries the governments of Europe and North America were actively involved in the management of natural resources. Of course, this was not management in the sense that we think of today. Rather, it was an effort to catalogue and exploit the great mineral and natural wealth available to the nations and their citizens. It involved support of expeditions to the interior—for example, the expedition of Meriwether Lewis and William Clark from the Missouri River basin to the Pacific Coast in 1804–1806, the charting of coasts and rivers, sounding of harbors, surveying of land and the timber remaining on it, and meteorological studies. Governments provided funds to growing networks of public and private universities for agricultural, silvicultural, and other research activities important to economic growth. Several countries also began to regulate resource exploitation in modest ways—through closure of forests, users' fees for access to water or grasslands, or taxes—to ensure conservation of resources. Sometimes governments acted in response to growing concerns about the health and safety of citizens—for example, through the construction of sewage and water supply systems in the late nineteenth century.

In the United States the Coastal Survey joined with the Army Corps of Engineers to bring about improvements in the nation's waterways and harbors. In Prussia foresters in service of the state advanced *Forstwissenschaft* to bring order and regular growth to the stands of trees. (The foresters recognized only belatedly that pruning and the clearing of the forest floor deprived the trees of essential nutrients for growth.) In England Parliament continued a long tradition of protecting the forests established under the crown to ensure access to materials needed for the navy. In Norway the government created a fisheries inspectorate in the mid-nineteenth century, to assist coastal fishing communities in maintaining their livelihood under the pressure of technological change.

Urged on by public health advocates, journalists, and naturalists, many of these states turned to regulation of industrial activity and attacked such problems of urban life as water pollution, waste disposal, and illnesses connected with them. Rapid industrialization in Europe and North America in the nineteenth century had had a devastating environmental impact. Seeking power for their mills, industrialists had dammed rivers and streams, felled forests for lumber, and mined coal without thought for the consequences. They may have assumed that resources were plentiful, or they may have worried that if they didn't take decisive action, their competitors would secure the resources before they did. The result was that by the end of the nineteenth century most New England streams had lost their salmon populations, many rivers contained dead zones, most forests near industrial population centers had been felled, and waste from mining, smelting, and burning of coal and wood filled the water and air. Tanning, textile, metallurgical, and food-processing factories often dumped their wastes into the nearest body of water. Once cities built sewage systems, the industries initially hooked into them only so that the waste was carried a bit farther away. The novel *The Jungle* (1906), in which Upton Sinclair shed light on the terrible filth and pollution of the meatpacking industries in Chicago, the squalor in which workers lived, and the dangerous conditions in which they worked, led to the passage of the Food and Drug Act of 1906.

While citizens generally believed that resources were inexhaustible, already by the mid-1850s and 1860s some naturalists had begun to worry about rapid depletion and about the extinction of birds and mammals. John James Audubon (1785–1851) contributed to early conservation by producing several multivolume books about all the known birds of North America. Henry David Thoreau (1817–1862), the essayist, poet, and philosopher, worried in *Walden: Or, Life in the Woods* and other writings about overuse of resources and the growing negative costs of "civilized" life. He criticized lumbermen for failing to understand the beauty of the pines the way a poet would.[2] George Perkins Marsh (1801–1882), author of *Man and Nature; Or Physical Geography as Modified by Human Action,* worried openly about the damage to New England ecology caused by lumbermen and farmers. He understood that human influence, though dynamic, was not necessarily beneficent and that the shortsighted treatment of the environment made human beings their own worst enemy.[3] On a walk from the Midwest to the Gulf of Mexico and on various trips in the Sierra Nevada in California, and to Alaska, Oregon, and Washington State, the naturalist John Muir (1838–1914) learned of the need for a federal forest conservation policy to protect the great natural treasures of the nation. As a preservationist, Muir hoped to set aside wilderness regions as inviolable, not to arrange for their managed use. His efforts helped force the federal government to establish national forests. Muir camped with President Theodore Roosevelt in the Yosemite region of California, an experience that influenced the president's embrace of

Progressive-Era conservation legislation and led to the establishment of national parks.[4]

THE RISE OF CONSERVATION MOVEMENTS

Pollution and the profligate use of resources led at the end of the
nineteenth century to the formation of conservation movements in
North America, Europe, and their colonial holdings. The movements often arose in connection with professional associations of
geologists, botanists, and biologists. To a degree, conservationists
accepted the view of such naturalists as Muir that resources needed
to be protected. But they also believed that through scientific management of those resources it was possible to use them rationally
and ensure their availability to future generations. In other words,
even if cattlemen, industrialists, small homesteaders, and a growing
number of vacationers all competed for the same resources in the
American West, scientific study would make possible "multiple use"
of resources for the foreseeable future.

As Samuel Hays argues, conservation was a "gospel of efficiency."
During the Progressive Era (around 1890–1914) in the United
States, scientists and engineers with different fields of specialization
sought to improve the quality of American life through science.
Science, they believed, offered the best way to achieve any given
goal, and generally it was applicable in all areas, from engineering
tasks to social organization. Scientists further believed that science
was an endeavor concerned with the objective truth. They could
streamline solutions to difficult social, political, and economic issues by providing policy makers with that truth, with that best way
to accomplish a given task. Professional organizations of scientists,
foresters, engineers, hydrologists, and others worked to establish
Progressive-Era legislation based on scientific facts; politics had no
call to interfere. Regarding conservation, specialists believed that
applied science could decide the course of resource development

and the appropriate distribution of wealth resulting from it, since resource-related matters "were basically technical in nature." Therefore, technicians and engineers, not legislators, should deal with these issues.[5]

Conservationists joined engineers in the effort to create science-based management of natural resources and to ensure "planned" and "efficient" progress—and not because of criticism of business practices that wasted those resources. Conservation neither arose "from a broad popular outcry, nor centered its fire primarily upon the private corporation," Hays writes. The conservationists were not pessimistic about the future out of some fear over rapid depletion of resources, but optimistic about the power of science to facilitate conservation goals. "They emphasized expansion, not retrenchment; possibilities not limitations," Hays writes. Further, many businesses welcomed a scientific assessment of the extent of natural resources, from water to forest to ore to wildlife, which might end cutthroat competition for them. Ultimately, Progressive-Era legislation set the stage for federal involvement in ensuring multiple use of the nation's resources, from recreation to extraction, from the establishment of parks to the felling of trees. Such legislation included the Newlands Act (1902), which gave the Reclamation Service, at first part of the U.S. Geological Survey, great power to build dams and irrigation systems and promote hydroelectricity, and the General Dams Act (1910), which fixed future licensing procedures. Legislation made possible the rational management of national forests, through the creation of the Forestry Service, in which Gifford Pinchot was a leading figure.

Gifford Pinchot proudly proclaimed the birth of scientific forestry. He asserted its ability to manage natural resources for generations to come. The science had been difficult to establish: since a forest crop takes decades to mature, it is difficult to gather empirically based observations on it. Experimental plots and stations were established only late in the nineteenth century, and government did not adequately support them. Laboratory work on the chemical composition, constituents, and possible uses of a tree and its secretions, along with the investigation of diseases and wood preservation, were all in their infancy. To expand investigations and secure scientific management of the forest, Pinchot sought the establishment of the Forestry Service, which in 1905 consolidated numerous different government activities and areas of concern (soils, hydrology—the study of streams, rivers, and lakes—weather, agriculture, wildlife, and so on) in one department.[6] Up to this time, several bureaucracies had competed for influence over forest resources and had failed to coordinate their activities.[7] The Forestry Service would solve all problems. Acting through it, the state would protect forests and streams from appropriation by private interests and from the rapacious extraction of coal, oil, iron, phosphate, and other minerals. Civilization, its cities, its "various mechanical aids to human life," all depended on the natural resources that provided the "raw materials of human existence." Pinchot concluded: "The conservation of natural resources is the key to the future. It is the key to the safety and prosperity of the American people, and of all the people of the world, for all time to come. The very existence of our Nation, and of all the rest, depends on conserving the resources which are the foundations of its life."[8] Since the Progressive Era, and not in the United States alone, representatives of government, industry, and the scientific community have come to share the belief that they can work together to make resources available to businesses at low cost (which would be passed along to consumers) and to serve citizens' interest in the conservation of resources.

In addition to their interest in forests, Progressive-Era specialists believed they had answers to the problem of inadequate water supply in the American West for irrigation, agriculture, and other purposes. The scientists needed only government funding to support their engineering skills in order to store rain and snowmelt runoff for later use. Seeing that the government supported canal, dam, harbor, flood-prevention, and road-building projects (what I call geoengineering projects) already in the East, Westerners sought funding for similar projects in the West. Eventually, Congress passed the Reclamation Act of June 17, 1902, which supported irrigation projects; users were to repay construction costs through various water-use fees. Supporters of reclamation argued that these government projects would encourage settlement and farming, to the benefit of all. Between 1902 and 1907, the Reclamation Service undertook more than two dozen projects in Western states. In 1923 the service was renamed the Bureau of Reclamation. Construction on the Boulder (Hoover) Dam on the Colorado River, begun in 1928, was crucial to the bureau on two levels. It was symbolic of the potential of nature-transformation projects to serve public and private interests, and it was a boon to bureau funding. During the Great Depression and until the 1980s, the Bureau of Reclamation was one of the major promoters of dam, canal, and irrigation projects in the world. The environmental movement of the 1970s and 1980s and the failure of the Grand Teton Dam in 1976 led to loss of the bureau's ability to command unquestioned authority and extensive resources.[9]

Pinchot and others credited President Theodore Roosevelt with launching the conservation movement. The various conservation acts of the Roosevelt administration enabled the government to protect the public interest, to prevent predatory, monopolistic companies from taking over the best sites for power development and then holding on to them without developing them, to hurt competitors. The government established charges for land use that were fair, including mineral and grazing fees.[10] Ultimately, the National Conservation Commission in 1908 brought together state officials to consider how to promote the best possible actions by the government to manage water, forest, land, and mineral resources. President Roosevelt said, "We intend to use these resources; but to so use them as to conserve them."[11]

Along with its involvement in managing resources, the government also began an innovative program to set aside land for recreation and preservation—national parks and monuments. A crucial event was the passage of the Yellowstone Act of 1872 establishing the first national park, preserving its land from settlement, occupancy, or sale, and setting it up as a public park "for the benefit and enjoyment of the people." The act gave exclusive control of the park to the secretary of the interior to set forth rules for the care and management of the park that provided "for the preservation, from injury or spoliation, of all timber, mineral deposits, natural curiosities, or wonders" within it. It was eighteen years before another park was set aside, and not until 1916 was the National Park Service established. But a precedent had been set.

In another effort to preserve wilderness areas, Congress passed the Antiquities Act in 1906. This act authorized the president to establish by public proclamation "landmarks . . . structures, and other objects of historic or scientific interest situated on lands owned or controlled by the United States government to be national monuments." It permitted such a designation for "the smallest area compatible with . . . proper care and management." (Congress also has the power to declare monuments, and has done so in twenty-nine cases.) Theodore Roosevelt, the first to hail the act, designated 18 national monuments in nine states. President Jimmy Carter declared fifty-six million acres in Alaska national monuments. In all, fourteen presidents used the act to proclaim 118 national monuments. The exceptions were Presidents Richard Nixon, Ronald Reagan, George H. W. Bush, and George W. Bush, his son, who vigorously attempted to reverse those designations, to enable mining and drilling on the land.[12]

Through legislation and the growth of bureaucracy, the secretary of the interior gained great power and responsibility regarding stewardship of vast tracts of federal land. But the process of balancing access to resources among competing interests is anything but scientific. It is a political decision to assign rights to recreation, grazing, lumbering, water, and other interests. In 1916, for example, Congress gave the secretary the power to sell or dispose of timber to control pests, stem the spread of diseases, or "otherwise conserve the scenery or the natural or historic objects in any such park, monument, or reservation." The law simultaneously permitted grazing of livestock "within any national park, monument, or reservation" except Yellowstone National Park. Finally, the law authorized the secretary "to construct, reconstruct, and improve roads and trails, inclusive of necessary bridges, in the national parks and monuments." The problem is that roads bring in heavy equipment, including for lumbering and mining activities. By 1999 there were 664,420 miles of roads on federal public lands, 446,020 of them under U.S. Forest Service jurisdiction, that had permitted access to, and clear-cutting of, forests.[13] A large dispute has also arisen over whether to permit highly polluting and noisy snowmobiles into Yellowstone National Park.

STATE-SPONSORED ENGINEERING OF NATURE

Hence, by the end of the nineteenth century, governments played an active role in nature management that went beyond cataloguing of resources, to include construction and transformation projects and regulatory functions. The construction projects comprised harbor improvement, direct and indirect involvement in the building of railroads, canals, and bridges, and management of rivers. The projects were often characterized by the hubris of conquest of nature. They also often had significant effects on the environment that were different from those anticipated.

A major actor in the engineering of nature in the United States is the Army Corps of Engineers. Todd Shallat writes that the corps's programs for river and harbor improvement rested on an American faith in progress through science and engineering, and also on an-

other kind of faith—that nature could be perfected. The corps, created in 1802 as a war academy and fort-building agency, became the
locus of multibillion-dollar projects to transform nature, largely toward the ends of controlling floods and bringing water to industry,
farmers, and municipalities. That is, like the Reclamation Service of
the Progressive Era, corps engineers fully subscribed to the Enlightenment ideal of the desirability of control over nature. Similarly, the corps was an engineering bureaucracy that demonstrated
clearly how the modern nation-state embraced science to make
improvements for the benefit of the citizenry. As such, the corps became the supporter of "extravagant" projects: dams, canals, aqueducts, lighthouses, levees, breakwaters, and ports. By the late twentieth century the corps had acquired fleets of 4,000-horsepower
tugboats capable of pushing fifty barges at a time and possessed
huge suction dredges, steam shovels, and other heavy earth-moving
equipment. Unfortunately, there is significant evidence that the
corps failed to understand that in its efforts to "engineer" nature for
the better, a task it took to be purely scientific, it engaged in policy
making—it usurped the power of Congress to select and advance
projects, and it overrode local interests.[14] Further, significant evidence shows that many projects led to irreversible destruction of
entire riverine, estuarine, and wetland ecosystems, while at the
same time leading to floods of greater and greater catastrophic impact.[15]

The first major corps projects centered on Mississippi valley
flood control. The engineers believed that if they built up levees
along the length of the river, they could channel water in times of
flooding into the Gulf of Mexico. They learned, albeit slowly and
incompletely, that levees merely channeled into the river greater
quantities of water, moving at higher speeds, which, in times of
flood, threatened the levees with scouring action. Levees became
weakened and waterlogged, in some cases giving way, with cata-

strophic impact. Since the levees were built, massive floods that in-undate hundreds of square miles have occurred with greater fre-quency.

It was not supposed to be this way. In 1879 the Mississippi River Commission was established to straighten and harness the river for navigation purposes (to keep the steamships moving), and also to make certain no water got out onto the floodplains. These plans illustrated perfectly the role of the state in sponsoring nature-management activities to promote the economic well-being of citi-zens and businesses. Many projects made good sense, for they de-fended significant property investment—for example, the St. Paul flood-control project to protect the downtown.[16] The efforts of the corps on the Mississippi were crucial for the timber industry across northern Minnesota and Wisconsin that provided white pine for a burgeoning housing and construction industry, and for the ship-ping industry to the south. Corps engineers were actively involved in pushing legislation to prohibit the dumping of such refuse as wood chips and sawdust into the river, which endangered naviga-tion efforts.

Flood control has remained the corps's major focus, as has the belief among the corps engineers that they could control floods. They convinced senators and congressional representatives—espe-cially those whose states included flood-prone areas—to give them substantial appropriations for their "aggressive effort to rebuild the Mississippi in their image: earnest, predictable, reliable." Yet the "levees only" flood-control strategy led the engineers to close sec-ondary channels and outlets and abandon upstream reservoirs that might have captured or dispersed floodwaters, and to rely instead on massive embankments separating the river from the floodplain. The projects encouraged the building of farms and towns on flood-plains, often on drained and reclaimed wetlands that had previ-ously served as a kind of sponge to soak up floodwaters. As a consequence, when floods occurred, considerably more valuable property became submerged and no wetlands remained to store water at times of peak flow and to maintain flow in dry times. Water that used to take weeks or months to move downstream now reaches the rivers in a matter of hours. Before the European colonization of North America there were about forty-five million acres of wetlands in the Mississippi River watershed, approximately 10 percent of the total area. Only approximately nineteen million acres of wetlands remain in the watershed.[17]

Engineering of rivers therefore turned out to be a disaster on two counts. First, it encouraged misplaced confidence that the river would no longer flood. This confidence led to more and more investment in towns and farms along the floodplains. Second, when the levees failed, they failed spectacularly. The 1927 flood along the lower Mississippi River displaced at least seven hundred thousand people and killed hundreds more. In some places the 1927 flood lasted half a year, destroying levees from Illinois to the Gulf of Mexico, inundating twenty-seven thousand square miles of land in all. To save New Orleans, the corps and state officials dynamited levees and flooded poor and politically disenfranchised parishes downstream. Federal involvement in disaster relief and recovery was rapid and massive, but white people fared significantly better than black people. The flood led corps engineers to abandon the "levees only" policy.[18]

In spite of efforts by the corps, serious, record-breaking Mississippi basin floods occurred again in the spring of 1973 and in the summer of 1993, in the latter case killing fifty people, displacing fifty thousand residents, and causing $12 billion in agricultural and property damage.[19] This has in no way dampened the enthusiasm of the Army Corps of Engineers, however. Huge, government-supported projects remained in vogue throughout the Western states, as the corps joined with the Bureau of Reclamation to tame streams and rivers through dredging, dams and irrigation, and canals. The projects created agricultural wonderlands through federally subsidized irrigation systems and provided hundreds of thousands of people with livelihood. A source of special pride was large-scale federal irrigation districts in the Columbia River basin, California's Central Valley (the nation's most important agricultural region), along the Snake River in Idaho, and along the North Platte, South Platte, Arkansas, and Colorado Rivers.[20] At the same time, the costly projects led to significant social displacement and the destruction of ecosystems, including the fish and wildlife in them. The finite amount of water available in the West made it a never-ending challenge to allocate it among agricultural, municipal, and other interests. To date, agriculture has garnered the lion's share of the water. But with the growth of cities in the West and Southwest, the system by which water is allocated has come under great stress.

Scholars, journalists, and eyewitnesses have written dozens of books on the history of the creation of hydroelectric power stations along the Columbia River, the most famous being the Grand Coulee Dam, the irrigation systems that the Bureau of Reclamation added to take advantage of vast amounts of water stored in great reservoirs, and the establishment of the Bonneville Power Administration (BPA) to administer the copious amounts of electricity now produced. The books cover the destruction of the lives and cultures of indigenous peoples in the flooded river basin, the annihilation of wild salmon, the failure of fish ladders, hatcheries, and other facilities to save the fish, and the siphoning-off of electricity for aluminum and plutonium production. Some have been written to glorify the great projects of the 1930s that helped extricate the nation from the Great Depression, and subsequent dams in the 1940s, 1950s, and on into the 1970s that contributed to the economic growth of the Pacific Northwest. Political leaders referred to the dams as evidence of what a democracy could accomplish even in times of crisis, and argued that the dams' benefits accrued to all citizens, as was fitting in a democracy, and not only to the chosen few. Others have lamented the irreversible costs of the transformation of the Columbia River basin into an "organic machine."[21]

For readers of this book, an important issue, beyond the effort to evaluate the costs and benefits of the expensive projects, is whether claims hold true that the construction of the Columbia dams typified projects under democratic institutions. For at the same time as Americans rebuilt the Columbia River, Soviet engineers were rebuilding the Volga River with a like number of huge hydropower stations. They did so in explicit and implicit competition with the United States, to demonstrate that socialism was better than capitalism. Soviet engineers and political leaders, no different from their American counterparts, claimed that their dams—for example, the mighty 2,000-MW (megawatt) Kuibyshev Dam on the Volga—in fact served the people, whereas American dams served the wealthy and supported the military establishment. That workers on Soviet projects endured much harsher living and working conditions, had higher injury rates, and were often imprisoned for failure to meet construction targets indicates that Soviet claims were exaggerated. Further, the benefits of Volga hydroelectricity went primarily to Moscow and served the political elite, but most of the electricity from the Columbia was distributed more widely; again, claims that dams are symbols of democracy seem to hold truer for the Grand Coulee.[22] Similar issues arise with Central Valley agricultural projects in California. Granted, irrigation water that the Bureau of Reclamation made available to farmers turned the valley into one of the most productive regions in the world. Yet agriculture has continued to receive a disproportionate share of the water available to Westerners, and the major beneficiaries have been huge agricultural businesses (hereafter "agribusinesses"), not ordinary citizens.

The government has also been centrally involved in the development of modern agriculture. State-sponsored research and development gave rise to soil science and knowledgeable use of fertilizers; modern hybrid crops; powerful machines for more efficient preparation, planting, and harvesting; stable production year after year, even in the face of fluctuations of rainfall and temperature; low commodity prices; and an abundance of goods. Yet awareness of how modern agricultural technologies have contributed to soil erosion, how chemicals, finding their way into streams and bays, have poisoned the land and led to eutrophication of water through algae blooms, and how huge machines have destroyed fragile ecosystems in the pursuit of farmland have led many citizens to worry about the environmental and societal costs of modern agriculture.

One of the most powerful arguments that capitalism was the root cause of the environmental crisis in the twentieth century can be found in Donald Worster's *Dust Bowl*. Worster examines the climatological, ecological, and social roots of the "dust bowl," areas of windblown erosion of topsoil in the plains states, from Texas and Oklahoma to North Dakota, in the mid-1930s, during an extended drought that destroyed agriculture. On several occasions the dust was so thick that it was dark at midday. Millions of tons of soil blew thousands of miles to the Atlantic Ocean. Yet was the dust bowl caused by unfortunate weather cycles alone? Worster points out how World War I and production problems in Europe created a boom for grain farmers in the United States. Taking advantage of low-interest loans and widely available land on the open plains, farmers used plows and tractors to break up the sod to grow grain. The plows destroyed the vegetation that had held the sod together. Businessmen followed onto the land, bought it up, and employed tenant farmers to raise grain for them. They released armies of tractors onto the land to prepare it for growing grain, and the technological preconditions for the dust bowl were in place.[23]

PUBLIC AUTHORITIES AS ACTORS ON THE ENVIRONMENT

In the United States and elsewhere, state-sponsored power, transportation, and agricultural organizations gained considerable authority in the twentieth century, especially during the Great Depression, as a response to unemployment, and during the cold war, in the pursuit of national security. In Canada such organizations include HydroQuebec, whose projects near Hudson Bay, detractors claim, have threatened the Cree Indians with extinction.[24] In France similar bureaucracies formed to support the multipurpose development of such river basins as the Rhône. In Germany, which had no actors as powerful as those in the United States or Canada, we could single out the Prussian economic ministry and later the Department of Transportation, which was active in building canals from the 1920s on, and the Rheinisch-Westfälisches Elektrizitätswerk (RWE), important in the history of hydroelectricity.[25] The Rhine-Main-Danube Company in the 1980s built the Rhine-Main-Danube Canal, which had first been considered in the 1920s.[26]

A symbol both of the significant contribution that state-sponsored large-scale projects can make to the public welfare and of the potential costs of that approach is the Tennessee Valley Authority (TVA) system of dams, reservoirs, and hydroelectric power stations. Gifford Pinchot asserted that TVA, grew out of the findings of the Inland Waterways Commission that reported to President Roosevelt.[27] The U.S. government created TVA in the 1930s, in part to pull the country out of economic depression. Federal projects produced jobs. But the TVA dams and reservoirs were more than job projects: TVA officials stressed the role of hydroelectricity in making the lives of rural inhabitants of the region easier and more productive. TVA sent out agricultural experts to teach new farming techniques and methods of soil conservation, such as contour plowing and crop rotation. As is true of many technological systems, hydroelectricity did not come alone but brought with it other large-scale systems. In TVA's case, this meant the use of nitrate and phosphate fertilizers produced at the Muscle Shoals, Alabama, facility. During World War II, TVA generating facilities, especially Watts Bar Dam and Fossil Plant, produced the copious amounts of electricity that allowed the United States to meet the wartime demand for aluminum production. In the 1960s and 1970s, TVA diversified into nuclear power, by building five reactors at three nuclear sites, in the belief that the nation needed to increase capacity and that nuclear power was cost-effective.[28] Even more undeniably, TVA produced democracy.

The promoters and directors of TVA stressed such benefits as flood control, recreation, and power generation. With good reason, supporters believed that electricity would illuminate—literally and figuratively—the hollows of Appalachia and raise the poor inhabitants inexorably out of poverty. Tennessee Valley Authority subsidized the purchase of new kitchen and laundry appliances that were seen as icons of public health and prosperity based on electricity. The agricultural programs of TVA also contributed to a significant increase in the quality of life. The first chairman of TVA, David Lilienthal, made explicit the connection between technology and democracy. Only in the United States, with its "dynamic decentralization," was administration of river resources for the benefit of all citizens possible. Only through reliance on citizen input could nations avoid the tyranny of technocracy. Lilienthal wrote about the central position of the citizen in TVA's plan: "People are the most important fact in resource development. Not only is the welfare and happiness of individuals its true purpose, but they are the means by which that development is accomplished; their genius, their energies and spirit are the instruments; it is not only 'for the people' but 'by the people.'"[29]

Water-supply projects that benefited urban residents also grew in popularity during this time. When, in the 1920s and 1930s, the demand for water in Boston and eastern Massachusetts grew, the state legislature approved funding to build the Quabbin Reservoir in the western part of the state. Hundreds of residents of four towns were evicted from their homes, houses were razed, cemeteries were moved (except for those of Native Americans), and a 104-square-kilometer (40-square-mile) reservoir filled a valley 100 miles (more than 60 kilometers) from Boston to serve that city.

Judging by the millions of people who benefited from TVA, BPA, and water district projects, these large-scale technological systems truly served the citizenry. Yet as with other large-scale approaches to the transformation of nature, there were also significant social and environmental costs. Indigenous peoples and the poor lost their homes; such anadromous (migrating) fish as salmon disappeared from many rivers and streams; big business and big agriculture were able to take advantage of heavy federal subsidies to push aside the small farmers who were intended to benefit. And the lion's share of electrical energy produced went not so much to the poor as to the burgeoning uranium separation facilities at Oak Ridge, Tennessee, to facilities at Hanford, Washington, producing plutonium for the atomic bomb, and to aluminum smelting for war machines. The radioactive waste and metallic sludge and other by-products of smelting are part of the environmental legacy of TVA and BPA.

For several reasons, little building of big dams takes place today in pluralist systems. The best reservoir sites have been developed. Cost-benefit ratios are adverse for most of the remaining sites. The public has become more sensitive to environmental and social issues. And the general perception is that big dams benefit agribusiness, not the family farmer, as was believed in the heyday of big dams. Especially in pluralist regimes, but also elsewhere, citizens have become much more effective in their opposition.

In peninsular Malaysia and in Tasmania, for example, large, undisturbed wilderness areas have disappeared as forests were felled, and as mining and roads encroached on them. Areas such as these often have endemic rare species that can survive only there. Developers then sought out wild rivers for hydroelectric power stations. Building the dams would require lumbering and road construction on, or flooding of, vast areas. In Malaysia officials selected a site on the Tembeling River. The Tembeling is the only path of travel for inhabitants in villages in the vicinity, whose Bronze-Age history would disappear under a 130-square-kilometer reservoir. The dam would require that 326 square kilometers be taken from a national park. In addition to loss of land and livelihood, the dams would lead to the production of hydrogen sulfide from decaying trees, which produces acids, along with rapid growth of water weeds, increasing silt, and fish kills. The Malayan Nature Society successfully defended the Taman Negara Park after a series of contentious public meetings. The government may have abandoned the project in 1983, as in other cases, not only because of public opposition, but because of growing cost estimates and declining projections of true productive capacity ($300 million for only 110 MW to provide less than 3 percent of the country's electrical power capacity). In Australia, after officials announced their intention to build a dam on the Gordon River below the junction with Franklin River in the heart of the southwest, public opposition eventually led to the abandonment of the project, but only after an Australian Supreme Court decision in 1985. Cases of successful public opposition have been the exception. Between 1945 and 1971 more than eight thousand dams were built around the world, many of them in North America, where in some regions virtually every river is dammed. In 1950s and 1960s the momentum to build dams carried over to Africa and especially to Southeast Asia and the Amazon basin.[30]

AQUACULTURE, SILVICULTURE, AND THE ENVIRONMENT

Specialists and industrialists have not been content with geoengineering projects. They have sought to turn dammed waterways into fisheries for sport and food and to transform specific flora and fauna into cash crops. These efforts grew naturally out of the increasingly systematic effort in agricultural research to hybridize corn and other crops and cattle and other animals. Already in the nineteenth century the engineering effort had focused on fish (aquaculture) and forests (silviculture). Aquaculture, largely in ponds, has existed on a small scale for hundreds of years in Europe and in China. Private hatcheries existed in the United States before the Civil War, and by 1875 several societies and associations formed to work with the U.S. government's Fish Commission to promote modern hatcheries. The products of inland hatcheries included trout, salmon, whitefish, walleye, bass, pike, and muskellunge. But the efforts of these groups to stock sport and food fish through hatcheries failed because of lack of understanding of the physiology and biology of fish, the release of fish into unsuitable habitats, and the inability of the fish to survive predators or compete with organisms present in the waters. The simple belief had grown out of the Enlightenment worldview that nature itself could be made industrial and that the transformation of biological organisms into highly productive monocultures was a reasonable path to follow.

Yet no science of fish ecology existed to consider the environmental issues attendant on the industrial fisheries. Few limnologists (freshwater scientists) or ichthyologists (fish specialists) were employed in state conservation departments. Also, fish farming was not profitable in North America until the early twentieth century, since commercial fishing on wild, unmanaged bodies of fresh water provided an adequate supply of fish to meet domestic demand. Once huge hydropower stations on rivers threatened fish populations by altering their habitat and preventing the migration of fish to spawning areas, fisheries scientists turned their attention to restocking rivers, developing fish ladders to permit migration past dams, and seeding fish in reservoirs. By the end of the twentieth century some local and regional officials, regulators, and energy producers asked whether the time had come to breach dams to permit anadromous fish to reach their upstream mating areas, for the costly fish ladders had failed to do anything to stem the decline of migrating fish populations.

Early aquaculture consisted largely of an effort to produce large numbers of fish from artificial hatcheries. How many of those fish survived after being transferred to rivers, lakes, and artificial bodies of water seemed less important. On paper, success had been achieved. In the water, however, few of the fingerlings lived to maturity. Fisheries scientists then realized that they also needed to study how to limit the growth of plants and algae, how and where to create better spawning conditions, when to regulate fluctuation in water levels, and how to prevent erosion and silting up of the waterways that engineers had "improved" with dams and irrigation systems, but where they had destroyed spawning areas. Hatcheries alone were inadequate. There needed simultaneously to be laws to protect native fish, closed seasons during spawning, redoubled efforts to stock artificially propagated and reared fish, and continued research.[31]

The connections between the state, technology, and the environment as regards the fishing industry are quite complex. State support for research was grudging at first and focused largely on data collection (about populations, flow of streams and rivers, rudimentary water chemistry, ocean currents, and temperatures). Little study was made of how to combat the impact of industrialization and hydroelectric projects on fish populations. Early efforts focused on restocking behind dams and in recreational fishing areas. Generally, scientists viewed both ocean stocks and those in inland bodies of water as an inexhaustible resource. Throughout Northern Europe and North America regulation was limited to local officials' requiring licenses for fishing on rivers, lakes, and ponds. By the 1920s many places had set seasons for fishing and limits on sites that might be fished. The impact of those regulations was small, however, and fish populations continued to dwindle.

Just before World War II, scientists began extensive pond research, particularly at Alabama Polytechnic Institute (now Auburn University). The facility consisted of more than one hundred ponds that could be drained and refilled and that ranged in size from one-tenth of an acre to more than one acre. Scientists elsewhere also studied fish habitats, food, behavior, and physiology. Scientists developed understanding of the role that fertilization plays and established that relatively small amounts of fertilizer might produce large numbers of fish. But most important was development of basic concepts that enabled scientists to vary characteristics of the ecosystem to judge the impact on fish. They learned that pond habitat restricted fish populations, such that the addition of more fish would not necessarily increase production beyond certain levels.[32] As the construction of reservoirs of tremendous size in connection with federally supported geoengineering projects gained momentum in the 1930s, scientists recognized that there was more to aquaculture than simple seeding—or releasing—and harvesting of fish. If they wished to seed fish into reservoirs, they had to know precisely which fish to seed and how to ensure their feeding and survival. Sometimes their efforts worked, and sometimes the fish died.[33] Experts found no simple answers or low-cost possibilities. State intervention in the form of regulations, users' fees, and fines was required to save fish populations and promote stable commercial activity. But no substitute existed for preservation of wild native stocks.

Electrical power stations provided opportunities for and obstacles to aquaculture. Regardless of whether power stations ran on nuclear energy or fossil fuel, they used copious amounts of cooling water from rivers or lakes, in some cases upwards of a million gallons of water per minute, and discharged it directly back into the water source below the point of intake. Some scientists believe that this practice may not create problems for certain indigenous species in winter or during flood stages. But during much of the year, when runoff waters and stream flow are low and seasonal temperatures are highest, the heated water may hurt fish and aquatic invertebrates. Builders try to avoid the costly solution of constructing large and expensive artificial cooling ponds as an interim stage before discharge. Some engineers ludicrously justified their hesitance to protect riverine and lacustrine flora and fauna from the discharge of heated water by arguing that the warm water stayed on top and had little effect on the layers beneath. This claim indicates that scientists were slow to recognize the extent to which the physical improvements they had introduced, such as dredging, straightening, and damming, along with thermal pollution, had significantly changed the physical and chemical characteristics of lakes and rivers. The suggestion to take advantage of the changed characteristics—for example, by introducing new kinds of fish downstream from heated effluents or building ladders for salmon to travel along, back to their spawning grounds—can be understood in this light.[34]

By the end of the twentieth century, large-scale aquacultural operations had developed for such anadromous fish as salmon. Promoters intended these operations to meet market demand and to make up for the decline in wild salmon populations caused by human predation, pollution, and dams on salmon runs from the Pacific Northwest and the Columbia River to Kamchatka, Russia, and from Maine to Canada, Greenland, Iceland, the United Kingdom, and the Norwegian coast. To reverse this trend, governments and businesses invested heavily in hatcheries to maintain the output of juvenile migrants into the oceans. Engineers maintain that hatcheries make it possible to eliminate many environmental mortalities (such as those from predation and pollution) and achieve production at levels ten to a hundred times that in nature.[35] In the case of Norway, the government supported salmon aquaculture to assist fishermen in coastal communities whose livelihood was threatened by huge factory trawler fishing. Salmon aquaculture has become widespread in Chile, Canada, Scotland, England, Iceland, Russia, the United States, and elsewhere.

Most consumers do not realize the extent to which farming of monocultures of flora and fauna has affected goods for sale at the supermarket. Customers understand that chickens, pigs, and cattle come increasingly from agribusinesses. Most people realize that packaged meats come from some distance away, and that the admonishment to "think globally, act locally" is no longer possible when it comes to food, clothing, and energy supply, all of which are produced at some distance from where Americans live. Some of them understand that the environmental and social costs of factory animal farming to produce "biomachines" (industrially produced animals) are substantial. A few recognize that most of the salmon they now purchase come from fish farms off the coasts of Russia, Norway, and the United States. Salmon aquaculture is relatively efficient, with a pound of pen feed yielding nearly a pound of salmon, but problems with fecal pollution, fish disease, and threats to natural species suggest that regulatory officials need to examine the future of aquaculture more closely.

Salmon ranching is based on the high degree of homing accuracy of salmon. When young fish are released to the sea through ponds, enclosures, or channels, they will return from marine waters to be harvested at a marketable size in those "homes." They can be caught as mature adults in restricted estuarine pen environments after swimming widely at sea; harvesting and processing take place onshore in local factories, followed by shipment to markets. Ranchers believe that enough breeding stock remains to prevent the creation of monocultures of salmon. Yet danger exists on at least two fronts. First, were the salmon to escape from their pens, they might "outcompete" wild salmon in the Darwinian sense. Many have been engineered to grow and fatten more rapidly than wild salmon, and since the females tend to spawn with the larger males, this is a danger. Second, the fish at many farms, especially in Maine, have already succumbed to diseases that threaten to wipe out entire farms—for example, a kind of anemia.[36]

STATE-SUPPORTED DEEP-SEA TRAWLING

Large-scale trawling fleets, complete with floating factories for processing and refrigeration at sea, destroyed ocean fisheries. After World War II, the nations of Northern Europe, the United States, Canada, and Russia used major technological advances in sonar and materials (such as strong, light plastics for nets) and the shipbuilding capacity from the war years to build trawling fleets capable of depleting entire fish populations in a matter of years. Often working in tandem, ships trailing drift nets kilometers long would pull on board hundreds of tons of fish for processing. The captains were not selective about the catch—they could not be—but hauled everything on board, dumping the dead fish they did not want.[37] They have essentially destroyed the cod fisheries of the North Atlantic; they have required closure of Georges Bank off the coast of New England, once one of the richest fisheries in the world; the fleets have overfished wherever they trawl. The government-subsidized and corporate-owned trawler fleets outcompete the smaller, privately owned fishing vessels of traditional coastal communities all over the world. Japan and the former Soviet Union were among the most rapacious of the trawling nations.[38]

The fishermen and women of Gloucester (Massachusetts), Newfoundland, Greenland, and Iceland have seen their livelihood disappear with the fish. Their local knowledge about where and when to fish has been overwhelmed by the universal knowledge of the trawling fleets.[39] Tim Smith argues that the desire to catch huge amounts of fish has created pressure on fisheries scientists to interpret fluctuations in fish catches and populations to favor continued fishing. That is, the notion of an "optimum catch" that will maintain fish populations developed under political and economic pressure.[40]

As for the oceans, government support for big trawling vessels through low-cost loans and subsidies and for research on the location of fish, in tandem with inadequate regulations, permitted big technology and big business to dominate the fisheries, as fishermen in coastal communities lost their livelihood. Their fleets could not compete. Regulators were slow to establish coastal limits to protect fish close to shore. By the time officials imposed limits on catch or shut down such areas as Georges Bank, the areas had been fished to exhaustion.

Among the reasons for the destruction of the North Atlantic cod fisheries are disengaged politicians, marine scientists who lacked understanding, and capitalist greed. Another explanation, one that grows out of the Enlightenment understanding of the human relation with nature, is connected with the vision that industrialization ought to occur in aquaculture no less than it did in agriculture, forestry, water management, and manufacturing. Visionaries saw industrialization and its attendant modernization as a means to transform traditional and "backward" coastal communities into components of the modern economic engine that included nature itself. People in those communities themselves were allegedly backward, working in several occupations, according to the season, resistant to change, and lacking innovative spirit. A modern, innovative state could encourage them to take advantage of the prosperity promised by capitalist industrialization. As elsewhere, the state in Newfoundland encouraged industrialization—in this case, of the fisheries—as a means of fighting the effects of the Great Depression. It pursued the larger economic and technological changes occurring in the fisheries as a solution. As Miriam Wright notes, industrialization was seen as "the true path to prosperity. It placed great faith in the ability of private capital, technology and rational state planning to create a better world."[41]

One of those technological changes was the decline of the salt fish trade that accompanied the rise of the frozen-food industry. Refrigeration technologies, especially rapid freezing technology that did little damage to the food, by permitting freezing without any formation of large crystals, enabled entrepreneurs to seek larger, more distant markets for various products. Growing urban centers offered precisely those markets. The salt fish trade tended to be local and small in scale, based on credit relations between fishermen and merchants. Frozen-fish technology relied on industrial organization of fishing trawlers and centralized processing plants and bought into fishing the profit-seeking and the dynamism of international markets—or so the visionaries believed, not without some justification.

In Newfoundland, during the Great Depression, British and Canadian civil servants sought to bring principles of efficiency and centralization to the fishing economy, to modernize it. They intervened to create a central bureaucracy to coordinate fishing, develop cooperative ventures, and revitalize the rural and coastal economy. On the eve of World War II, the government offered low-interest loans for the construction and operation of refrigeration plants and the purchase of deep-sea trawlers. The government actively recruited salt cod merchants to make the transition to frozen fish. Many of them took advantage of loans to build modern trawlers and refrigerated processing vessels. Between 1946 and 1964 some two dozen frozen-fish plants came into operation in Newfoundland. When Newfoundland joined the Canadian federation in 1949, the new federal government, in concert with provincial officials, continued to advance modern fisheries technologies, in pursuit of a partnership among state, private capital, and industry. The belief that technology helped the Allies win the war played no small role. The federal government now promoted through advertisements and pamphlets many of those technologies—radar, electronics, and trawlers—which stood in stark contrast to the backward fishing traditions of coastal communities.[42]

Competition for fish in the North Atlantic intensified in the 1960s, as a dozen nations trawled far and wide for cod and other fish. The Canadian government, like the others, determined to support the effort to expand trawling, in part to promote economic development of Newfoundland and the Maritime Provinces, which lagged behind urban, industrial Ontario and the grain farms of the prairies. But like the other nations that fished the North Atlantic, Canada belatedly recognized that wreckless fishing, which modern technology had made possible, would destroy the fisheries. In 1977 Canada joined other nations in extending its national limit to two hundred miles offshore. It came too late for the fisheries. But the government never lost faith in technology to solve the problems of the collapsing fisheries. Rather, when officials saw European nations catching fish with modern trawlers, their response was to encourage the same. The government expanded Canadian programs to allow manning of larger trawlers and fleets equipped with modern gear to track down cod and other fish. Finally, in July 1992, the federal government of Canada was forced to close commercial cod fishing on the east coast of Newfoundland, on the Grand Banks, and off the coast of Labrador; the next year other sites closed as well. Overnight, over thirty thousand fishermen and plant workers in Newfoundland became unemployed.[43]

In several cases, small-scale resource management practices based on traditional ways of life persisted into the late twentieth century. One is the Norwegian herring industry, where fisherman-owned firms rather than corporate or national fleets have predominated. The state helped preserve this industry by regulating entry into the fisheries, providing low-interest loans through a state bank, and giving fishermen control over the market. When such industries as herring and salmon fishing take place on a small scale, investments to replace worn-out or broken equipment (drift nets, traps, weirs, and rods) do not significantly decrease income, because successful fishing depends more on skill and labor than on equipment. The take being small, the number of suitable fishing sites is also limited, and this in itself prevented expansion or over-fishing. But once technological development outstripped the contribution of local skills and labor, two sets of things happened. First, the sites for fishing increased in number, the catch grew larger, and there was more competition for fewer fish. Second, local skills and local labor became less important, local income declined, and absentee owners of fishing fleets began to push the fishermen aside.

Between 1956 and 1962 there was a sharp decline of the Norwegian herring stock, perhaps the result of overfishing. In any event, fisherman-owners continued to fish, even with the drastic drop in profitability, while nonfisherman-owners turned to other fisheries or quit fishing, leaving the small firms with a vested interest in the community to survive. Ethnographer Cato Waldel writes, "The hazardous nature of the fishery, with its periodic drastic drops in returns, results in a withdrawal of the financially strongest and largest firms and helps maintain the prevalence of fisherman-firms." A parliamentary committee on ownership rights with regard to fishing vessels concluded, "The people who pursue their societal duties by fishing in bad as well as good times have a right to protection from society against speculation by capital-strong individuals and companies in good times."[44]

To maintain the vitality of fishing communities under threat from technological change, the Norwegian government promoted regional development of other sectors of the economy where fishermen might augment their income. Officials hoped that the agricultural sector would produce supplemental income through salmon aquaculture.

The freshwater fisheries—including salmon—fell under the jurisdiction of the Ministry of Agriculture, while the saltwater fisheries had their "own" Ministry of Fisheries. The involvement of the state in promoting salmon aquaculture in fishing communities was understandable; nature was still an unpredictable factor, so an individual who invested exclusively in one sector might lose everything. Yet because of that state involvement, individuals were required to have knowledge of matters beyond agriculture or fishing, let alone aquaculture, for to make intelligent decisions about generating supplemental income or securing loans for farm supplies, fishermen had to be able to maneuver through the state subsidy and tax structure. They had to understand the rules of the game, and those the rules were made outside the community, in the parliament in Oslo. Local fishermen's associations that had regulated activities, distributed licenses and sites, and encouraged cooperation in such areas as trap-weir-fishing had to learn these rules, too.[45]

To spread the risk over more activities, Norwegian policy makers also explored a combination of Atlantic salmon and trout farming aquaculture with traditional fishing. The parliament funded aquaculture, for local municipalities remain dependent on it to keep employment levels sufficiently high year round. There are now roughly three thousand aquaculture sites along the Norwegian coast, employing twenty-three thousand persons in fleets, thirteen thousand in processing, and forty-five hundred in fish and shellfish farming. Aquaculture in 1997 produced 316,000 tons of salmon and 34,000 of trout, or a total catch of 2.8 million tons of fish, 95 percent of it for export, making Norway the tenth-largest fish producer in the world.[46]

Is aquaculture the best way to ensure that local fishing communities remain healthy in the face of seasonal fluctuations, external economic pressures from other kinds of fisheries, and difficulties in competing with large-scale ocean-trawling industries? Or is aquaculture, with its monocultures of fish, its government subsidies, and its industrial ethos a large-scale, environmentally disruptive technology itself, even in such places as Norway where government officials and scientists make the effort, when promoting modern science and technology, to consider local concerns? In promoting aquaculture, technological development, and traditional fisheries simultaneously, Norwegian policy makers and fishermen ran head on into the problem of how to balance local, traditional knowledge with state- and science-based knowledge. Could the parliament support industrial aquaculture without hurting traditional fishing villages? Svein Jentoft, Connor Bailey, and others have pointed out that fisheries management is a complex problem, one aspect being the difficulty in developing consensus on what constitutes relevant knowledge and facts, owing to scientific uncertainty.

Can a democratic government balance the interests of local fishing communities with those of industries, in the face of unrelenting technological change? Are there scientific facts on which to base policy decisions? Or, as in the case of Progressive-Era policy makers and engineers who sought to manage water and forest resources scientifically, is it difficult if not impossible to avoid mixing decisions about resource management and issues of scarcity with politics? In democratic societies different interest groups attempt to influence the policy process, and therefore the "facts," whereas in authoritarian regimes, lobbying, manipulation, and political pressure are of necessity kept to a minimum. At first glance, in other words, better knowledge will not necessarily ensure the best policy in democratic societies, whereas in authoritarian systems it may.[47]

IS AGRICULTURE ANY DIFFERENT IN ITS SOCIAL AND ENVIRONMENTAL IMPACTS?

The support of modern agricultural techniques and technologies by modern nation-states has also had paradoxical effects on society and environment. On the one hand, modern Western agriculture produces a cornucopia of products year round at very low cost. On the other hand, government support for agribusiness has contributed to the destruction of the small family farm, the concentration of production among a few major producers, and such undesirable environmental degradation as erosion and water pollution.[48]

It was not intended to be this way. Government-sponsored research and development were supposed to lead inexorably to rural self-sufficiency and to production at levels adequate to supply growing urban markets. The Morrill Act, which the U.S. Congress passed in 1862, made possible the establishment of state-sponsored universities and land grant colleges. The goals included extending education to more and more Americans, not only in Eastern and Southern cities with established schools, but in areas of settlement in the Midwest and West. A grant of federal land to each state would enable states to sell land and use the proceeds to establish colleges in engineering, agriculture, and military science. More than seventy land grant colleges were established by 1890. They developed from leading agricultural and technical schools into excellent public universities generally. They received direct support from the U.S. Department of Agriculture for agricultural research.[49] The extension services of the colleges and universities provided knowhow to struggling farmers. They conducted extensive research with federal and state funds on experimental plots to hybridize plants and develop irrigation systems, fertilizers, and insecticides. Yet increasingly the research effort benefited big businesses through new cash crops and mechanized planting and harvest that only wealthy farmers and ultimately agribusinesses could afford. Over time, paradoxically, the U.S. Department of Agriculture research programs therefore hurt the small farmer they were intended to support.

The close connection between big business, scientific research, and government gradually resulted in consolidation of farms into big businesses. This has been the trend since the late nineteenth century.[50] Clearly, the achievements of modern research are remarkable: diverse crops, more of which make it from field to supermarket, expanded production, a huge export market. Yet the family farmer has disappeared. Leland Swenson points out that neither record-low crop prices, natural disasters, nor economic recessions abroad but rather concentration into agribusinesses was the greatest cause of the decline of small farms. In the United States this concentration has left almost all sectors of U.S. agriculture in the hands of a few well-connected megacorporations. By 1999 four firms controlled 80 percent of the cattle slaughter business, 60 percent of the pork-packing industry, over half of chicken slaughtering, 74 percent of corn production, 62 percent of wheat production, and so on. The number of hog farms declined from 580,000 in 1981 to 139,000 in 1997. Large production facilities (those with more than a thousand hogs) account for 5.9 percent of operations but 63 percent of total inventory. Vertical integration permits a few firms to control the entire market, from production to processing and marketing.[51]

While these farms may be the epitome of planned, rational industrial animal production, they have hidden costs. First, as Swenson writes, concentration "has significantly contributed to a decay of community infrastructure in many rural areas." When local firms and banks are involved, they are interested in providing jobs and investing profits in the local community. Multinational corporations, which make profits for distant shareholders, rarely show any interest in local investment opportunities.[52] Second, from the environmental perspective, concentration has been a blight. According to a study by the Sierra Club, giant corporate-owned factory farms, so-called Concentrated Animal Feeding Operations (CAFOs) have become major polluters whose activities remain largely unregulated. According to the Sierra Club, CAFOs "produce staggering amounts of animal waste . . . (2.7 trillion pounds per year). Too often, this waste leaks into our rivers and streams, fouling our air, contaminating our drinking water and spreading disease. According to the Environmental Protection Agency, hog, chicken and cattle waste has polluted 35,000 miles of rivers in 22 states and contaminated groundwater in 17 states."[53] Small farms much more efficiently reuse manure and other wastes. Third, the industry suffers not only from concentration and pervasive, unsolved pollution problems, but from significant worker safety problems and the production of foods that are often tainted with life-threatening bacteria.[54] Concentration was the result of trends typical for both capitalism and Enlightenment thinking about the desirability of pursuing industrialization of biological processes.

Another manifestation of the joining together of Enlightenment vision, government support, and corporate innovation is the creation of genetically engineered plants and animals (genetically modified organisms, or GMOs). Increasingly standard products of agribusinesses, GMOs have become widespread especially in the United States, where the Food and Drug Administration and the Department of Agriculture have faced pressure to approve them rapidly as safe and efficacious. Produced by such corporations as Monsanto, the products involve xenotransplants (transplants of genetic material of bacteria and animals from one species into another), which may increase yields, protect against infestations, reduce frost damage, make possible mechanical picking and packing, and postpone ripening until produce (such as Calgene's Flavr Savr tomato) reaches the stores. The companies that produce GMOs tout the efficiency and predictability of their products, and the ability to control them from planting to purchase. Many crops respond to specific biocides at specific times. Soybeans, for example, have been engineered to tolerate applications of Monsanto Corporation's Roundup herbicide for use against weeds and grass.[55]

Yet throughout Europe, the United States, New Zealand, and Australia, opposition to GMOs has grown. Many people believe that the organisms will be dangerous to human health and the environment in ways we cannot yet predict. The worry is that there are not enough checks on the development of GMOs and that corporations will underestimate the risks because of the tremendous profit potential of GMOs. In some African countries, even though

GMOs promise crops with higher yields in difficult growing conditions of low rainfall and poor soil, citizens have rejected the modified seed. Some people argued that "natural" plants are more democratic, for anyone can grow them. They are naturally more diverse
and freer of "genetic pollution." Still other critics were angered that
patented seeds tied farmers like indentured servants to corporations and kept people from producing their own seeds from crops.
Of course, humans have been modifying natural plants and animals
for millennia through selective breeding and hybridization, so what
is natural and what is dangerous have become contentious subjects.

NUCLEAR POWER AND THE ENVIRONMENT

One of the most intractable environmental problems of the twentieth century concerns the vast and growing quantities of radioactive
waste. This waste is a serious problem for two reasons. First, much
of it is highly toxic. Second, while some of it consists of short-lived
radioisotopes, much radioactive waste is long-lived, lasting well
over ten thousand years. Determining what constitutes safe storage
or disposal under these circumstances has truly been a difficult
task, one that no nuclear nation has solved, and it is an epic problem, owing to the presence of 350 nuclear power stations worldwide in several dozen countries. Few cases resemble that of radioactive waste, where specialists knew full well from the start the extent
of the handling and disposal problems yet did little to meet the
challenge, beyond establishing "temporary storage" facilities that
are now decades old, overburdened, decrepit, and leaky.

Radioactive waste is formed when uranium ore is milled to produce fissile fuel. Only .7% of the uranium found naturally is fissile.
In the separation of fissile from nonfissile isotopes, vast amounts of
waste are formed, much of which is radioactively contaminated. In
addition, so-called plutonium production reactors have been used
to transmute nonfissile U^{238} into plutonium for nuclear weapons.
The uranium (and the plutonium produced from it) are encased in

steel-clad fuel rods. Huge concrete "canyons" filled with acids and other chemicals are then used to dissolve the cladding. Tens of thousands of tons and millions of liters of low- and high-level radioactive waste have been produced in nuclear nations through the weapons fuel fabrication processes, none of it (as of 2002) in permanent storage. Medical treatments, experimental reactors, and food irradiation (cold sterilization) also produce large amounts of radioactive waste.

The two main producers of radioactive waste were the United States and the former USSR. Hiding behind proclamations of "national security interests" and assuming they would eventually figure out how safely to dispose of radioactive waste, American and Soviet engineers put it in tanks above- and belowground that soon leaked dangerous radioactivity into the groundwater. The slurries of acids and radionuclides were poured into ponds and other impoundments that gave way or leaked. At the Hanford Atomic Reservation in Washington State, the site of several plutonium production reactors, engineers oversaw the creation of vast quantities of these wastes, which have been making their way into the groundwater, and thence into the Columbia River—and its salmon—for decades.[56] To make matters worse, between 1945 and 1963 France, China, the USSR, the United Kingdom, and the United States detonated hundreds of nuclear weapons into the atmosphere that spread radioactive fallout over the globe, with significant detriment to public health in the form of increased numbers of cancers and genetic mutations. Public outcry brought an end to atmospheric testing of weapons only in 1963.

While the production of nuclear weapons has had the most serious deleterious impact on the environment, the dispersal of civilian power reactors (for electrical energy) has also exacted very high costs. Since the late 1950s, civilian nuclear power plants have stored spent fuel rods on-site, mostly in pools specially constructed to keep the rods cool and stable. Having run out of space, stations are increasingly storing the spent fuel rods on-site aboveground in so-called dry-cask storage. For example, since it began operation in 1972, Maine Yankee in Wiscasset, Maine, has produced 1,434 used nuclear fuel assemblies. These fuel assemblies are high-level waste. It is the responsibility of the U.S. Department of Energy to dispose of this waste. Because the department does not have a disposal facility available for the waste, Maine Yankee stores it on a picturesque peninsula in a tidal basin, where it will remain at least until 2023 in aboveground air-cooled concrete casks.[57] In 2003 more than one hundred civilian power reactors were in operation in the United States, most located in densely populated regions along the east coast, in Illinois, and in California, which store spent fuel rods in "temporary facilities," some of them decades old and full.

In 2002 the U.S. government finally approved a permanent radioactive waste disposal site deep under Yucca Mountain, not far from Las Vegas, Nevada.[58] In fixing on a disposal site, government and industry scientists were guided by several considerations. One was that the depository should be in a region of low seismic activity. Another was that the excavated geological site must be dry. Salt domes are appropriate in this regard. Yet another criterion is that storage must be reversible in some sense. That is, it should be secure enough to keep unauthorized persons out, but accessible to others who might have to remove the waste if advances in engineering suggest an alternative in hundreds or thousands of years, or in case the site fails (because of seismic activity, moisture in and corrosion of casks, or terrorist activity). A final problem is the transportation of waste to any repository. What if a major accident occurred? Freight trains and trucks carrying casks of radwaste have to be able to withstand high-speed, high-impact crashes and fires. There will be tens of thousands of shipments, all of which will have to be secure. These dangers notwithstanding, many people support the construction of new nuclear power stations (and therefore the production of more fuel and the generation of more waste) as the only solution to the growing energy crisis in the industrial world.

Nuclear power has a significant presence in the energy balances of France (80 percent), Hungary (40 percent), Great Britain (23 percent), the United States (20 percent) and many other countries.[59] The advantages of nuclear power are clear. Reactors do not contribute such greenhouse gases as carbon dioxide to global warming, do not emit oxides of nitrogen or sulfur or particulates, and do not contribute to emphysema or other lung diseases. In the process of mining and milling uranium, much less waste is produced, and less environmental damage is done, than in coal mining. Significantly fewer deaths and diseases are tied to nuclear power generation than to fossil fuel. Still, nuclear power engenders great opposition because of the intractable problem of waste and the potential for a catastrophic accident. Siting processing facilities and reactors far from population centers might be a way to deal with these concerns, but then transmission costs would increase. Finally, reactors produce tremendous amounts of heat, which has significant local impact in the form of effluents from reactor-cooling water that reenters lakes, rivers, or the ocean and thereby threatens local flora and fauna. Nuclear power stations require vast areas of land, both to accommodate the reactors themselves and service buildings and to serve as an exclusion area to keep the public out. And the human and environmental costs must be weighed, from the increased levels of lung cancer among Navajo uranium miners to exposure of workers to excessive radiation during milling, handling, and transfer of nuclear materials and during operation of the plants.

Of course, the greatest concern from the environmental standpoint is the possibility of a catastrophic nuclear accident that would eject the products of fission far and wide or lead to a meltdown of the reactor core. Nuclear power stations tend to be located near population centers to keep transmission and associated capital costs down. Consequently, were a serious accident to occur, tens and hundreds of thousands of people might be exposed to dangerously high levels of radioactivity. Recent studies by state and local emergency and nuclear plant personnel contradict those of the federal government's Nuclear Regulatory Commission, which maintain that evacuation plans to ensure the safe removal of those people from the contaminated areas are viable. (Recall the massive traffic jams when people flee hurricanes—but in those cases with days of the warning to evacuate, as opposed to minutes.)

In several pluralist regimes—Sweden, Germany, and the United States, for example—public opposition to nuclear power led to the closure of stations, or in essence a moratorium on construction. The European Union has insisted that Lithuania close its Ignalina station as a precondition for joining the European Union. In the United States, the public has been directly involved in assessing the viability of this technology. The Clamshell Alliance, an NGO in New Hampshire, temporarily halted construction of the Seabrook Nuclear Power Station in its tracks. One reactor was eventually brought on-line, even in the face of clear evidence that no workable evacuation plan could be put in place for residents in the surrounding area in case of an accident. The other planned reactor was never completed. In general, in the U.S. protest against nuclear power has led to greater scrutiny of the safety and licensing procedures for nuclear power stations, and to requirements for retrofitting old stations and redesigning new stations. The result is that costs of nuclear reactors have gone up three- to fourfold, making them noncompetitive with fossil-fuel boilers and leading essentially to a moratorium on their construction in the United States, even as other nations continue actively to pursue nuclear power. These nations include Japan, Taiwan, Korea, France, and Russia.

A large number of serious nuclear accidents have taken place. The two best known were the one at Three Mile Island, Pennsylvania, in 1981 that led to a partial core meltdown and the one at Chernobyl in 1986, in which a massive explosion spread radioactivity throughout the northern hemisphere and required the establishment of an exclusion zone thirty kilometers in diameter. Dozens of other accidents have nearly caused extensive loss of human life and significant environmental contamination. In one case in Detroit, engineers suffering from hubris built a breeder reactor, based on an unproven technology, which in 1966 melted down, nearly requiring the evacuation of tens of thousands of people. On March 22, 1975, electricians trying to seal air leaks at TVA's Brown's Ferry Nuclear Power Station mistakenly set fire to foam insulation with the candles they were using to test for air leaks. The fire alarm, which they turned on belatedly, disabled such devices as an emergency core-cooling system and nearly resulted in a meltdown of the reactor's core.[60]

In spite of the problems of waste and the potential for catastrophic accident, isn't nuclear power a good alternative to continued reliance on coal and oil? Fossil fuels have long been the major source of air pollution. Regulation of air pollution in the twentieth century was ultimately a success story, although more remains to be done. It has been difficult to wean manufacturers and power producers away from coal, largely because of its relatively low cost. One turn-of-the-century observer went further, calling coal "fundamental to the civilization of the present era" and lauding its ability to help nations recover, with a burst of economic growth, from the loss of their forests. The high level of civilization in Great Britain rested on coal, this engineer observed. By contrast, Italy, with its limited resources, had a population "of low intelligence and low standards of living." The cause was not "the degeneracy of the Latin stock" but an economic crisis, at the root of which was inadequate power production.[61]

The nations of Europe and North America have for decades been aware of the need to regulate pollution. But their efforts to control or preclude air, water, and land pollution lagged until the 1960s. As noted earlier, from its early decades the Army Corps of Engineers promoted legislation to prohibit dumping in waterways. In European countries, the 1960s awareness of the dangers of fallout and chemical biocides served as a wake-up call. In Ireland, officials noted the presence of acid deposits on buildings and streets (and thus in citizens' lungs!) from a variety of pollutants, most produced by combustion processes and emissions, and including sulfur dioxide and nitrogen oxides. Short-term exposure to elevated concentrations was damaging—if not more so than long-term exposure to lesser concentrations. Yet before an October 1964 law on planning and development, the only air pollution regulations in Ireland were connected with the "Alkali etc. Works Regulation of 1906" and some provisions of the Public Health Act of 1878. These codes were entirely insufficient to meet the challenges of modern industry and automobiles and carried such vanishingly small penalties and fines as to be meaningless. Officials did not act, even as problems with smog grew, because outside Cork, Dublin, and other industrial cities Ireland had the good fortune to enjoy prevailing winds that carried the filth away.[62]

To deal with the growing threat of air pollution, the U.S. Bureau of Mines of the Department of the Interior created a (short-lived) Office of Air Pollution at the beginning of the twentieth century. This was an acknowledgement that the switchover of stoves, ranges, house-heating boilers and furnaces to bituminous coal, from the cleaner-burning anthracite for which they had been intended, had led to significant smoke problems. Since America depended on coal for heat, power, manufacturing, and railroads, the smoke grew quite dense in many locales. Also, too many businesses used their facilities inefficiently—for example, in the production of coke, by allowing the gases to escape, rather than trapping and tapping them for fuel. Citizens increasingly demanded smoke abatement, and not only in the United States but in European countries. Efforts commenced with public ordinances requiring that new plants be properly equipped and old ones refurbished, and that permits be issued for installation of boilers and furnaces. Dozens of American cities had passed ordinances by 1908. But most ordinances were ineffective, either explicitly exempting existing manufacturing plants or permitting higher densities of smoke than they ought, or opening loopholes by their vague wording. All provided penalties for emission of "dense black smoke" from furnaces. One specialist asserted that technological advances could solve the problem of smoke, for the causes were "lack of air at the proper temperature at the point where the volatile gases should be burned," which resulted in incomplete burning. Insufficient draft and inadequately trained operators who opened and tended furnaces too frequently also contributed to the problem.[63]

During the late 1940s serious smog incidents in Los Angeles and in Donora, Pennsylvania, raised concerns about air pollution. The city of Pittsburgh was infamous for its cloudy darkness at midday. Worse still, in London, a killer smog that blanketed the city from December 5 to December 9, 1952, killed at least four thousand people. In 1955 the U.S. government passed the Air Pollution Control Act, a law that ultimately lacked the scope and enforcement powers to be effective. The publication of Silent Spring, followed by growing public awareness of the dangers of unregulated pollution and by Earth Day in May 1970, led to the promulgation of the Clean Air Act of 1970, a successful and effective law, though not without its faults and detractors, which was still in effect with modifications in 2003.

The Clean Air Act strictly regulated emissions from new sources and assumed that many older polluters would eventually close or incorporate modern pollution-control technologies. It also set standards for controlling hazardous emissions and automobile exhaust and achieved significant successes in both regards. The act codified the right of citizens to pursue legal action against any person or any organization, including the government, that violated the emissions standards. In 1977, Congress amended the act to extend deadlines for automobile emissions. The Clean Air Act of 1990 gave states the power to establish their own deadlines for meeting targets and tightened automobile standards. The new act acknowledged the need to fight acid rain, by encouraging the use of low-sulfur coal and mandating the installation of the Best Available Control Technology (BACT) in any new facilities.[64] The original Clean Air Act had probably failed in its initial version by establishing so-called air quality regions throughout the nation, rather than national standards. This choice failed to address the fact that air pollution knows no boundaries. It also gave states responsibilities, but no means of enforcement. And not a single state developed a full pollution-control program as stipulated by law. That industry opposed the Clean Air Act nevertheless indicates that it had teeth. Yet efforts have persisted to free coal from a "morass of federal, state and local regulations" that hampered coal development, with supporters of coal pointing out that reserves in the United States would provide energy independence for centuries if industry were permitted to avoid using scrubbers, to strip-mine, and to gain access to coal on federal lands.[65] The quality of the air has improved greatly around the nation, and it has become clear that the executive branch must be willing to pursue enforcement of laws if air quality is to continue to improve.[66]

Other successes include Irish efforts to develop more efficient fossil fuel technologies for heating homes and business. In Ireland after the 1979 world oil crisis smog problems were exacerbated when many Irish homes began to burn bituminous coal. Efforts to convert existing systems to smokeless ones lagged, as the government undertook a survey to see how many houses in some places burned coal—knowing beforehand that almost all did. Finally, the legislators passed the Air Pollution Act of 1987 to make up for deficiencies and meet increased demands from the public and the

European Community for a cleaner environment. The law defined polluters and pollutants and gave power to the Ministry of the Environment to make regulations, set policy and standards, coordinate the activities of local authorities, and establish much larger fines. The law reflected European Economic Community laws and the 1979 Convention on Long-Range Transboundary Air Pollution.[67] This legislation showed that policy makers recognized how many kinds of pollution move through air and water, across boundaries. Democracies, more than closed, authoritarian regimes, are sensitive to transboundary pollution issues.

INDUSTRIAL POLLUTION AND HAZARDOUS WASTE: A "LUXURY" OF INDUSTRIALIZED DEMOCRACIES?

The successes that the state has achieved in regulating potentially dangerous industrial activities—forcing cleanup of hazardous waste, providing access to clean water, and confronting the excesses of technological know-how—indicate the positive role it plays in mitigating many of the negative aspects of modern industrial life. Some observers complain that the state possesses the political power, scientific knowledge, and financial wherewithal to achieve much more in the way of creating regulatory structures that function satisfactorily. Others worry about misguided efforts in some nations to erode that power in the name of protecting individual property rights and free-market economic activity. But most citizens in pluralist regimes welcome the achievements of the state in fighting the legacy of pollution, facilitating environmentally sound economic growth, and protecting the environment for use by their children's children.

In Jonathan Harr's *A Civil Action* (and a movie of the same title), the lawyer-protagonist attempts to identify the parent company responsible for dumping highly carcinogenic wastes near residential areas of Woburn, Massachusetts, that have led, the lawyer maintains, to increased rates of leukemia in the area. The connection between the wastes and leukemia was difficult to prove, but eventually the W. R. Grace Company had to pay substantial damages for its role in the disaster. The investigative and legal proceedings indicate how difficult it is to prove liability, let alone identify and punish polluters.

Another case of successful state intervention concerns the town of Love Canal near Buffalo, New York, whose residents suffered from a series of maladies connected with exposure to hazardous wastes. Love Canal was a seventy-acre site that included a sixteen-acre hazardous-waste landfill and a forty-acre clay/synthetic liner cap. William T. Love had originally excavated Love Canal in the 1890s for a proposed hydroelectric power station, which was never built. According to the U.S. Environmental Protection Agency, "beginning in 1942, the landfill was used by Hooker Chemicals and Plastics (now Occidental Chemical Corporation) for the disposal of over 21,000 tons of various chemical wastes, including halogenated organics, pesticides, chlorobenzenes and dioxin. Dumping ceased in 1952, and, in 1953, the landfill was covered and deeded to the Niagara Falls Board of Education."[68] Subsequently, developers turned the area near the landfill into a town with homes and schools. In the 1960s and 1970s, the hazardous wastes contaminated the water and land near those homes and public health suffered. Although Love Canal presents a horrific story of corporate obfuscation and delay, in the end, largely because of citizens' involvement, President Jimmy Carter ordered all residents of Love Canal moved from their homes into safe housing. Only because of that federal intervention were the citizens protected.

In the United States, Superfund legislation in 1980 led to the establishment of a trust fund, fed by taxes on industry, to clean up some of the worst hazardous waste dumps, such as those at Woburn and Love Canal, no matter what their origin. By then, many of the original polluters had gone out of business and others had been forced to declare bankruptcy to avoid the high cost of cleanup and liability. In the 1970s and 1980s workers dressed from head to toe in protective gear were commonly involved in recovery of the waste and its proper disposal at Superfund sites. The reasonable thinking was that the waste was too dangerous to permit us to lose any time establishing ownership or liability for it, and that American citizens deserved an environment safe from teratogenic and mutagenic substances (that is, those causing defects in the embryo or fetus). An important feature of environmental regulation in democratic countries is the assumption that ensuring citizens' health and safety must be paramount.

Dealing with the legacy of hazardous waste requires political will, in addition to money. The legal framework for municipal and industrial hazardous waste in the United States was the Resource Recovery and Conservation Act (RCRA) of 1976, its amendment in 1984, and CERCLA—the Comprehensive Environmental Response, Compensation, and Liability Act (1980)—the Superfund act. These acts concern solids, liquids, sludge, powders, or slurry wastes that are toxic, infectious, radioactive, corrosive, reactive, or ignitable. They include dioxins, asbestos, and PCBs, and other well-known poisons; explosives, plastics, refuse, inorganic pigments, plating and polish, wood-preserving chemicals, and many other substances. About 90 percent of them are in liquid form, and many are not degradable. Unfortunately, the Superfund act has not slowed the generation of additional hazardous wastes, which continues unabated.[69]

Further, the Superfund laws ignore pre-1940 sites. One author estimates that 5.7 million tons of hazardous waste were disposed of in 1935 alone (by metal and chemical producers and leather tanneries). These have neither deteriorated nor become diluted over time. What of lead used in smelting, batteries, ammunition, paint, piano wire, plumbing, and printing supplies; chlorine in paint and lacquer; waste from the rubber and dye industries; arsenic in glass, pesticides, and waste from chemical and metallurgical processes?

While industries often recommended funding for research on how to recover usable wastes from various processes, if no such techniques were found, they abandoned the waste and sometimes diverted industrial sewage to municipal treatment plants.[70] There are more than sixteen thousand waste sites in the United States, alone, but little documentation exists about what was dumped, where, and when.

Industry in the United States produces hundreds of millions of tons of wastes annually, excluding radioactive wastes—85 percent of the world's total hazardous waste—and there is no inventory to tell us how much or what kinds. The chemical, petrochemical, and metallurgical, facilities in the nation number in the thousands. Industry produces more than three tons of waste per person per year in the United States, and the number of sites, estimated at between two thousand and ten thousand, that need treatment under CERCLA indicates that the Superfund is underfunded.[71] The result is an industrial legacy of hazardous materials whose cleanup and proper disposal will cost hundreds billions of dollars. Who is responsible for the cost of the cleanup?

The Superfund has paid for work on 30 percent of the sites that have already been cleaned up, and corporations have paid for the rest. Beginning in the 1990s the U.S. Congress failed to agree on reauthorizing the tax that funds the Superfund. The corporate tax expired in 1995, and the fund shrank from $3.8 billion in 1996 to $28 million in 2002. The Bush administration, which opposed reauthorization of the requirement that polluters pay to fund the Superfund program, instead asked taxpayers pay for 54 percent of the Superfund program's costs. The administration of George W. Bush determined to clean up fewer sites through use of general government revenues, meanwhile easing the tax burden on corporations at sites where the responsible party cannot be identified or is unable to pay.[72] Are there any inexpensive or nonstatutory means to clean up hazardous waste? Should citizens be responsible for cleanup of industrial waste?

Legislation to clean up waste, regulate its production, and fine polluters is not enough. Other northern-tier nations have sought to deal with waste in more proactive ways, by reducing the amount produced and ensuring local participation in the policy process. There are four basic ways to deal with waste: through minimization of its production; recovery or recycling; biological, chemical, or physical treatment; and disposal. If disposal is the choice, waste can be incinerated, reused as fuel, solidified or otherwise stabilized for burial, impounded on the surface, placed in landfill, injected underground, or treated (for example, in a wastewater treatment facility). Nuclear engineers at one time suggested using underground peaceful nuclear explosions to burn waste. In the throwaway society of the United States, policy makers have chosen disposal. Paper remains the major source of discards, but plastics, which are not biodegradable, make up an increasingly large share of the waste—witness the replacement of paper with plastic bags in supermarkets and the ubiquitousness of plastic bottles for water. Could Americans do without so much packaging?

Sanitary landfills that at one time served as a "universally accepted disposal panacea" were developed by the Army Corps of Engineers and adopted by the Sanitary Engineering Division of the American Society of Civil Engineering in the 1950s. As the cost of land increased and local governments banded together to prevent the location of landfills in their communities, the landfill became a less attractive option. In addition, many of these facilities have filled up and been closed. There were roughly 2,200 landfills in the United States in 2000, with ten or so more opening each year. Incineration may be an option for waste, but most locales oppose having incinerators built in their communities. Recycling will certainly be needed to reduce the amount of waste accumulating.[73]

The quintessential landfill—the largest in the world—was Fresh Kills, a 2,100-acre site several hundred feet high in spots, on Staten Island, New York. When it opened in 1947, it was billed, like nuclear waste facilities, as "temporary," but it operated until 2001. Because it lacked the liners of more modern landfills, it leached toxic chemicals and heavy metals into the groundwater. When the Verrazano Narrows Bridge opened in the 1960s, Staten Island began to draw settlers from other New York boroughs, who began to protest the odor, environmental dangers, and spurious legality of the facility. By 1986 the landfill was receiving nearly thirty thousand tons of garbage from New York every day. Once the facility closed, another problem arose for residents of New York: Where could they ship their garbage? Dumping in the ocean was no longer an option; no neighborhood wanted an incinerator; and many out-of-state landfills had closed. Another problem—what to do with Fresh Kills— was solved with its reclamation and transformation from "landfill into landscape"—meadows, parks, and trails.[74]

Even in pluralist regimes, much of the burden of waste disposal falls on poor communities, the majority of whose residents often are people of color. This inequity contributes to what is known as environmental racism. Environmental racism exists in wealthy countries with democratic institutions that are supposed to distribute the risks and benefits of living in modern society equitably, just as it exists in developing nations. (And as I discuss at greater length in Chapter 3, wealthy countries can afford to deal with the problems of hazardous waste, sometimes by exporting that waste, especially to African nations.) There is some debate whether negative effects of governmental actions to regulate environment have been purposely concentrated on the more disadvantaged members of society. Recent studies by such scholars as Robert Bullard indicate that not poverty alone, but race, was the crucial predictor of these costs. People of color are more likely to live in counties where hazardous-waste disposal facilities are located.[75]

According to another observer, poor and minority communities appear to be exposed to larger numbers of polluting facilities and higher levels of pollution. These exposures and facilities are the result of market forces, local land-use regulations, and environmental permit decisions (for example, the cost of property in middle- and upper-middle-class neighborhoods makes siting of waste facilities of any sort in them prohibitively expensive). Does environmental racism violate the Fourteenth Amendment to the U.S. Constitution, which ensures equal protection under the law? Evidence exists that fines for violating environmental statutes are higher in white areas than in minority ones, and in wealthy areas than in poor ones. With respect to court decisions and environmental equity, federal courts seem to have systematically "encouraged" pollution—or discouraged pollution abatement—in poor neighborhoods by levying fines for polluting that are 50 percent lower than those in white areas and in wealthy areas.[76]

Countries with democratic regimes have—in part because most of them are blessed with access to copious amounts of water, in part because of their engineering achievements of reservoirs, canals, and pumping stations to secure it cheaply year round, and in part because of the promulgation of a legal framework to protect water resources and force the cleanup of pollution—succeeded in making access to clean water a right, not a privilege. Industrial production, suburban sprawl, and continued assault on wetlands through reclamation may threaten that right. For example, as Adam Rome writes, the postwar construction boom in the United States was responsible for the rapid installation of septic systems for wastewater disposal. They were low in cost but often hurriedly built and poorly designed. They effectively postponed decisions about proper waste disposal until the creation of sewage treatment facilities and burdened homeowners with costly repairs and retrofitting, and municipal governments with the high capital costs of those treatment facilities.[77]

The alternative to septic systems is either on-site wastewater collection systems that are tied into a public sewer system or independent on-site collection and treatment systems. Both systems use biological treatment in combination with oxidation ditches, rotating biological contractors, trickling filters, and activated biological filters. They typically remove 85 percent or more of organics and solids in the water. They also disinfect the water before discharging it directly into surface waters. But the best route may be conservation, which can reduce load by two-thirds or more through such simple devices as showerhead restricters and water-saving toilets (25 percent) and by separation of black water from gray water (another 40 percent) at the house, after which the gray water can be used for gardening, cleaning, and so on. Water conservation can significantly reduce the amount of wastewater generated, and thereby not only the costs of water itself, but day-to-day costs.[78]

In pluralist systems that sanctify individual property rights, citizens welcome the status and comfort that accompanies home ownership. Laws and policies encourage ownership, through low-interest loans and tax deductions for interest. Yet the industries that arose to meet demand for housing and the transportation and other technological systems that developed around houses have their own, as yet underestimated, costs. For example, the development of suburbia in America has had environmental consequences beyond those of wastewater treatment. Beltways divorce the city from the suburbs and cut swaths of development through the countryside. The absence of sidewalks and bicycle paths encourages homeowners to clog the streets with automobiles. A building boom in the postwar years exacerbated these problems when developers rapidly erected tens of millions of individual houses. Homes usually had washer and dryer, garbage disposal, air conditioner, and other energy- and resource-intensive devices. Developers focused on short-term benefits of low-cost construction that made houses cheap to buy. But they have been expensive to heat and cool, as federal, state and local building codes have only grudgingly required insulation, thermopane windows, and other energy-saving technologies. In addition, the homes are inordinately large by international standards, with the result that Americans require much more energy than citizens of other countries to keep their homes warm in winter and cool in summer. Ought not a democratic government to establish excise taxes to encourage modest home designs, and building codes to ensure their environmental friendliness?

Further, developers chose to clear land completely of vegetation, and it was often necessary to truck topsoil back in when construction was over. The open land was vulnerable to mudslides. Trees that had provided shade or wind protection disappeared under bulldozers. The lawns that consumers planted bolstered a $30-billion-a-year seed and fertilizer industry. According to Adam Rome, this approach to home construction in fact engendered an environmental movement, for Americans worried about the degradation they saw in and around their homes much more than they did about pollution far away. The NIMBY (not-in-my-backyard) attitude grew up on the unsafe water that came through taps, the soapsuds on rivers, streams, and lakes, the clearing of forests, the absence of parks, and the rise of suburban sprawl, and that attitude led to the passage of the National Environmental Protection Act, the Clean Air Act and the Clean Water Act.[79]

AUTOMOBILES, POLLUTION, AND THE STATE

A key to the rise of suburbia and what has come to be called sprawl was the automobile. The history of automobile pollution provides an example of the complex relation between technology, environment, and the state. The problem with the automobile, as with many other technologies, was that its costs were significantly greater than initially gauged. They included dangerous exhaust emissions, the requirement to build extensive systems of support (repair facilities, service stations, refineries, and research into and development and exploitation of reserves), and a reduction in the resources available for public transportation, as funds were diverted to road construction. Automobiles contribute more to greenhouse gas production in the United States than any other source, and they have become a major source of pollution in virtually all nations.

In each nation that has adopted the path of rapid industrialization, the impact on the environment has been roughly the same: industry provides many jobs but pollutes extensively; increasingly intensive agriculture puts pressure on the soil, requires heavy applications of chemical fertilizers, and accelerates migration to cities that themselves become extensive sources of pollution and human misery; and public health measures and environmental protection enforcement lag. In short, the embrace of the internal-combustion engine—in particular, the automobile—has accelerated environmental degradation.

In all systems considered here—pluralist, authoritarian, and postcolonial—people consider the automobile a right, not a privilege. First in North America and then Western Europe, the automobile quickly became a mass consumer item, not merely a toy of the wealthy or political elites. In authoritarian and colonial regimes, ownership of an automobile remained largely the province of citizens who were well connected or wealthy. Yet the automobile everywhere was simultaneously a symbol of power and prestige, of individual rights, and of modernity.

The older the automobile, the more heavily it pollutes, because it lacks such equipment as a catalytic converter and is often in bad repair. Smoke-spewing, inefficient older models that use lead fuel have migrated to Central and South American nations that lack the political will and legal structures to regulate them. Similarly, as soon as the Berlin Wall fell, West Europeans sold their older, polluting automobiles to Eastern Europe and the former Soviet Union. Overnight, throughout Eastern Europe, cities began to resemble parking lots. Rush hours lasted from morning to night. Leaders of the newly independent nations were unwilling to regulate the automobile, because the citizenry, so new to democratic institutions, viewed any regulation as a return to the authoritarian rule of the recent past. Instead, the new governments used tax revenues to build new roads, which encouraged the purchase of still more automobiles, which required still more construction, and so on, in a vicious technological circle.

The policies of municipal and national governments encourage automobile pollution in all its manifestations. Rather than provide an adequate level of support for public transportation, government instead accommodated the automobile, and expenses for highway construction, bridge repair, and creation of parking lots and monstrous parking garages skyrocketed. The pollution of air (smog) and water and hazardous waste associated with vehicles continued to grow, while rates for heart and lung disease, including emphysema and asthma, increased, as did deaths and injuries from accidents. Oil, brake fluid, and battery acid were discarded haphazardly by unthinking automobile owners and unscrupulous repair shop managers.

The automobile commands considerable power in Africa, too. South Africa, the most industrialized country of Africa, must battle extensive smog. Air pollution combines with water and chemical pollution of the gold, diamond, coal, and chrome industries to create dangerous living and working conditions. One of apartheid's legacies is that many black workers travel dozens of miles to work, then back to homelands. The homelands are on the worst land, with degraded soil. The commute in smoke-belching old automobiles exacerbates the pollution.[80]

No matter where it was adopted, the automobile took its toll. In Iran, beginning in the 1950s with the nationalization of the oil industry under Prime Minister Mohammad Mossadegh, development programs were launched to transform the nation from an agrarian into an industrial economy. Cement factories, oil refineries, construction companies, incinerators, smelters, and meat-processing facilities were built everywhere. Industrialization, along with rapid population growth, led to increased urbanization, a higher standard of living, and increased consumption levels. The rivers that flow north into the Caspian Sea and south into the Persian Gulf became filled with waste, garbage, and pollution. When oil prices rose, many newly rich Iranians bought automobiles. This led to congestion and air pollution, including extensive amounts of lead, nitrous oxide, and sulfur oxides.

Iran's Islamic revolution in 1979 slowed economic growth significantly, and economic stagnation eased pollution problems somewhat. But according to official estimates, the Iran-Iraq War triggered extensive health and environmental problems for 12.5 million Iranians, destroyed 5 million hectares of agricultural land, 8 million hectares of forest, and 8.5 million hectares of pastureland.[81] In 1984 the Islamic Republic passed environmental legislation that, on paper, looked progressive. It was intended to prohibit pollution of all Iranian waters and maintain existing air pollution laws. Iran sent representatives to the 1992 Rio summit, supported environmental research, and encouraged public awareness of relevant environmental issues. Yet the automobile remained a major polluter, and public transport was inadequate to ease the problem.[82]

Governments have addressed the problem of automobile pollution with varying degrees of success. Some of the solutions have been technological. In many countries, governments have required manufacturers to adopt catalytic converters to limit some of the pollution. In others, governments have passed laws to increase the gas mileage that vehicles must achieve. In the United States, for example, the Corporate Average Fuel Economy (CAFE) standards were established by Congress in the 1980s. Fleet averages (the total number of passenger vehicles sold in a year) were to rise to 40 miles per gallon by 2000. But the CAFE standard of 27.5 miles per gallon has not been increased since the 1986 model year, and manufacturers have used special designations for vehicles (*SUVs, light trucks,* and so on) that are not considered passenger vehicles, and thus do not have to meet the CAFE or safety standards, in order to build and sell heavy, gas-guzzling machines. The failure of the federal government to provide leadership has led state governments to adopt their own approaches. One example is California, where all automobiles sold must meet emissions standards that are stricter than those provided for by federal law. California claims to be the fifth-largest economy in the world. Forty percent of its greenhouse emissions come from automobiles. Governor Gray Davis signed legislation in July 2002 (which was opposed by the automobile industry) that required fleet average reductions in emissions to almost zero by 2009. The law did not impose new taxes, gasoline mileage limits, or any special restrictions on such vehicles as SUVs, but only the requirement that technology—available technology, such as electric cars—be used to lower emissions.[83] After initially opposing the legislation, the manufacturers have agreed they will try to meet the challenge.

In addition to technological solutions, governments can also use economic and political coercion. Throughout Europe, governments employed taxes to raise the price of gasoline by the equivalent of seventy-five cents to a dollar per liter, to encourage the use of bicycles and public transportation and pay for that infrastructure. In Trondheim, Norway, and other municipalities, the authorities lowered the speed limit to thirty kilometers (eighteen miles) per hour and built speed bumps to "calm" and discourage traffic. Such other municipalities as São Paulo and Mexico City have had to establish "no-driving days." The Mexico City metropolitan area has nearly three million vehicles, which average nine years of age; only models built since 1991 have catalytic converters. Fifty percent of all emissions from mobile sources are produced by private automobiles; 25 percent come from public transportation, and the rest from trucks. The pollution got so bad that the government introduced no-driving days in November 1989. The program required motorists not to drive on one workday during the week in phase 1 and not to drive on two days during the week (weekends included) in phase 2. Initially, the program was successful, netting a 20 percent reduction in the number of automobiles on the road, an increase in traffic speed, and a decline in gasoline consumption. The authorities decided to make the program permanent. However, once the program became permanent, drivers sought alternatives. Many found public transport uncomfortable and so purchased a second vehicle, in order always to have at least one vehicle available on any given day.

Finally, certain solutions to automobile pollution are political. Several of these can be undertaken domestically, for example, building well-subsidized public transportation systems—subways, commuter trains, trams, and buses. Unfortunately, public transportation requires extensive public investment, and southern-tier nations can rarely afford it. Even the United States is unwilling to subsidize public transportation to the level needed. For example, while the national passenger rail company, Amtrak, has received $30 billion over thirty years, automobile and air transport have received well over $1 trillion in the same period. International political solutions to pollution also exist. They include the Kyoto accord, by which the nations of the world have agreed to reduce their greenhouse gas emissions. But, as discussed in the concluding chapter, it remains uncertain whether Kyoto will come into force.

THE PENDULUM OF ENVIRONMENTAL PROTECTION MOVEMENTS

What role do environmental protection movements play in the development of environmentally friendly technologies and effective policies and laws? Movements for nature conservation and preservation came into existence to challenge the rapid pace of environmental change and degradation brought about by modern industry. By the beginning of the twentieth century such movements existed throughout Europe and North America. They often grew out of professional organizations of biologists, geologists, foresters, and others. The scars the industrial revolution had left on the landscape horrified many of them.

As historian Raymond Dominick shows, in Germany a NIMBY movement had already developed by 1900. German citizens fought a plan to erect a foundry in a well-to-do Hamburg suburb; they sought to protect the scenic rock outcroppings in the Ahr River Valley from quarrying; the Bürgermeisters of Berlin protested the felling of local forests and the sacrifice of remaining open spaces. Scientific societies (geologists, ornithologists, and botanists) with an interest in nature; professional associations of foresters and fishermen; public health groups (medical authorities, city planners, and government officials) who worried about city planning, potable water, epidemics, and sewerage disposal; and members of outdoor clubs all contributed to the nascent nature movements. Soon organizations specifically devoted to the protection of the environment were formed, such as the Bavarian State Committee for the Care of Nature. Its members were deeply disturbed about the damage to the Isar River and the Walchensee from a proposed hydroelectric power station (eventually completed in 1923). Others included the German League for Bird Protection and the Nature Park Society. By 1918 conservation societies numbered more than fifty thousand members all told.[84] These clubs were similar to the Sierra Club and the Audubon Society in the United States.

During the Weimar Republic (1919–1933), conservation movements continued to thrive in Germany. Article 150 of the Weimar constitution declared, "The monuments of history and of nature as well as the countryside enjoy the protection and care of the state." As the economy recovered from the devastation of World War I and inflation never before seen until 1923, the need for nature protection grew, especially given the problems of brown coal strip-mining, water pollution, draining and destruction of groundwater, and careless dumping of overburden. Some of the conservation groups lost membership and influence, and total membership dropped to forty thousand in 1925, but then grew again steadily until World War II. (As is discussed briefly in Chapter 2 on authoritarian regimes, little conservation legislation was passed during the Weimar period, a fact that enabled the Nazis to attract the conservationists, Dominick tells us.[85])

In the 1970s and 1980s, a Green Party developed in Germany, as traditional productive sectors of the economy (agriculture and manufacturing) declined. As in the United States, many members of the German Green movement were younger persons who rejected elitist decision making for grassroots participation, and some of them also resisted society's growing materialism and desired "a more simple and satisfying lifestyle," Dominick writes. They turned to a third political party because, as in the United States, it was difficult for voters to distinguish between the two main parties, which drifted toward empty messages and the middle ground, with a consequent lack of true legislative representation, let alone meaningful laws.[86] In Germany, electoral laws allow candidates of smaller parties easy access to the ballot, whereas in the United States the two-party system is dominated by the Republicans and the Democrats, who make it very difficult, through antiquated laws, for other candidates to get on the ballots. Also, in Germany any party that receives 5 percent of vote will have a representative in the Bundestag (parliament), whereas in the United States you must win the election in order to gain a seat. In 1983 the German Green Party won twenty-seven seats in the national legislature,[87] and it continues to wield influence. It is crucial to observe that Green parties and environmental movements have been most prevalent in democratic countries, and almost nonexistent elsewhere (see Chapter 2).

Another contributor to the formation of a vital environmental movement was concern about the dangers associated with nuclear weapons. Fears centered on the devastation of human life, of course, but also on the increasing levels of radioactive fallout that had found its way into the food chain, including into mothers' milk, and such potential outcomes as "nuclear winter," which would result if nuclear explosions created vast clouds of dust that filled the atmosphere and led to rapid cooling of the earth's surface. Such organizations as Pugwash and International Physicians for the Prevention of Nuclear War urged citizens to protest the growing danger of fallout. Many of the organizations were antiestablishment, and many were tied in with other movements (antiwar, women's rights, antipoverty, or civil rights). They successfully lobbied for the Nuclear Test Ban Treaty of 1963, which prohibited detonations of nuclear devices in the ocean, atmosphere, and outer space.

Environmental movements around the world gained much of their impetus from the publication of Rachel Carson's *Silent Spring* (first in serial form in the *New Yorker* in 1962 and then as a book early the next year). In *Silent Spring* Carson, a marine biologist, exhaustively documented the unthinking overuse of chemical poisons—herbicides and pesticides—and their entry into the food chain. She pointed out that these poisons had mutagenic and teratogenic effects and had killed birds, fish, and mammals outright. The biocides had grown out of World War II chemical weapons programs. By 1960 U.S. industry was producing nearly 300,000,000 kilograms per year, a fivefold increase in thirteen years—1.6 kilograms of these poisons per person per year.[88] To fight such seeming pests as gypsy moths, state and local governments had engaged in trigger-happy eradication programs. Crop dusters sprayed millions of acres of land with DDT. The chemical led to the acute poisonings or deaths of dozens of people and raised the question of who was the target. By 1986 more than four hundred insects and mites had become resistant to insecticide.

Carson did not reject pest control but sought to replace chemicals with biological methods. The response to Carson's book was prompt and furious. Representatives of the chemical industry sought to portray Carson as an emotional female and certainly not a scientist whose findings held any validity. But President John F. Kennedy ordered the formation of a commission to look into the conclusions of *Silent Spring*. Officials in the Kennedy administration and the general reading public came to the conclusion that at the very least little reliable information was available about the costs and benefits of the chemicalization of American life. States did not keep systematic data on exposures, illnesses, and deaths. Genetic research was sorely lacking, as was evaluation of the pathways chlorinated hydrocarbons took into plants, animals, and the food chain. As if that weren't bad enough, clouds of automobile exhaust or detergent suds marred most vistas.

Over the next six years, Americans experienced increasing consternation over environmental damage worldwide, including what they saw on television in reports on the war in Vietnam, which revealed the use of chemical defoliants and bombs to destroy the landscape. In an atmosphere of protest and growing concern, the U.S. Congress passed the National Environmental Protection Act of 1969. President Richard M. Nixon signed the bill into law, and the Environmental Protection Agency (EPA) itself came into existence in 1971. The Clean Air Act (1970), with its statutory deadlines for reducing particulates, carbon monoxide, nitrous oxide, and other pollutants, and the Clean Water Act (1972, amended in 1977), with its stipulations to clean up waterways and reduce effluent emissions, followed in short order.

The environmental movement was also invigorated by the study *Limits to Growth* (1972), which attacked the view that economic growth could continue indefinitely and, through the trickle-down effect, eventually benefit the poor. Profligate use of resources, a growing population, and deteriorating environmental conditions suggested that business as usual in North American and West European countries had to come to a stop. Citizens' groups mobilized to fight what the promoters of big projects advertised as progress. They managed to thwart highway projects and create networks to protect scenic rivers, and then they turned their attention to nuclear power. In the latter case, they lost out in major confrontations (examples are the Clamshell Alliance, which attacked the cost and environmental risks of the Seabrook Nuclear Power Station in New Hampshire, and the Abalone Alliance which fought against the Diablo Canyon Nuclear Plant, built along a seismic fault on the California coast).[89] Yet the Sierra Club, the World Wildlife Federation, the Audubon Society, and dozens of other conservation groups with extensive membership in many different countries indicate how vital are broad-based public interest groups to the operation of democracies.

Public awareness has been a critical factor in encouraging new approaches to waste management. The nations involved in the 1992 international summit in Rio de Janeiro on sustainability and efficient resource management—based on decentralized implementation at the local level—adopted Agenda 21 to promote those ends (for more discussion of Rio, see Chapter 4). While many of the European nations (such as Germany, Denmark, and Sweden) have embarked successfully on implementation of various programs, in the United States little has happened. All of Sweden's 288 municipalities pursued practical implementation of Agenda 21 to deal with hazardous waste. Sweden has a relatively long history of legal codes designed to deal with waste, but did not attack the problem with great vigor until the 1960s. (Its 1876 act on arsenic regulated handling of and trade in chemicals.) Politicization and regulation of hazardous waste commenced in the 1960s with the growing awareness of the dangers of biocides. Regarding mercury, for example, in early 1960s ornithologists pointed to evidence that birds were being destroyed. The source of the problem lay in the treatment of seed with mercury, in particular that used to kill fungal bunt disease, which affects grain during spring sowing. Swedish scientists had played a major role in developing this mercury treatment. Later, scientists also found mercury in fish, most likely stemming from chemical and paper industries' pollution of the lakes and waterways. The government established new regulations to limit discharges of mercury. By 1990 the government had arrived at a decision to eliminate the use of mercury (and other substances) through deep underground disposal. The problem with mercury is that it is fluid at normal temperatures, volatile, and of high toxicity. Mercury can never be decomposed or destroyed. So the effort to dispose of it properly for ten thousand years—let alone forever—was quite difficult.[90]

German officials focused on reducing production of waste, and not by shifting the burden of disposal to the consumer, but through laws that require manufacturers to be responsible for packaging at every step of the process. This requirement led the manufacturers to seek efficiencies in production and packaging that they might otherwise have ignored. This policy gained impetus from the reunification of Germany in 1992 because the perceived shortage of landfill capacities became a reality with the closing of several major sites in the former (so-called) German Democratic Republic. Germany generated about 150 million tons of solid waste annually, with residuals from construction, demolition, and excavation adding another 120 million tons. Sixty-five percent of that amount arose from production, and the remainder from consumption. German policy makers considered the waste from production a minor problem, since it results in large homogenous quantities and the firms involved in waste generation have to bear the responsibility for disposal. Elected officials believe that taxes on waste disposal that reflect environmental damage and the capacity of landfills will remove many of the inefficiencies in "the allocation of resources between waste prevention, re-use, recycling, and disposal."[91] Waste associated with consumption was more difficult to control, since such waste was dispersed among millions of consumers, heterogeneous and uncontrolled. The growing popularity of pay-as-you-throw programs that charged households for individual amounts indicated some success in promoting waste reduction. But pay-as-you-throw was limited in its effectiveness to returnable, as opposed to throwaway, containers. So the German government turned to "product stewardship," by making producers responsible for their products from cradle to grave, which led the producers to find the cheapest ways. The 1986 law on waste management was the legal basis for this German program. Now 80 percent of the materials used in packaging must be recycled. Significant implementation problems arose, to be sure, including licensing fees that turned out to be too low. The system has fared well, however, leading officials to conclude that good regulations can induce industries to find ways to pollute less.[92] You can now see color-coded recycling bins on street corners throughout Germany, and citizens use religiously separate their glass, paper, and cans quite efficiently. Would such a system work in the United States? Would mandatory curbside recycling cut down on garbage disposal? Are the costs of such a system worth the benefits? And what of placing some of the onus on packagers? Should, say, fast-food restaurants rather than consumers be responsible for the waste those products generate?

THE CONTINUING DEBATES OVER USE OF NATURAL RESOURCES

Many people argue that environmental protection laws constitute an unnecessary brake on the economy, reducing its dynamism and curbing the innovativeness of entrepreneurs who fear costly efforts to comply with laws. The simple fact that some of the largest and most dynamic economies in the world have some of the most vital environmental legislation indicates that this argument must be examined more fully. In addition, many leading corporations have taken the lead in promoting "green" manufacturing procedures, for they recognize that their long-term costs will be reduced.

Some economists and technology assessment experts argue that the cost of saving lives through regulation of technology or limiting exposure of humans to carcinogenic and teratogenic substances is so great as not to be worth the expense. Such critics point out that life in a modern industrial society carries some risks—for example, those connected with driving an automobile—which cannot be fully avoided. Other critics note that the food supply is safer than ever before in human history, so that the costs of entirely eliminating the risks of, say, a few additional cancers because of the presence of minute quantities of chemical residue on foodstuffs, far outweigh the benefits of the very few lives saved. Several skeptics have estimated the cost of regulation to save a single human life in the billions of dollars, depending on the regulation.

Yet these estimates, of "mythic proportions," according to Liza Heinzerling, were popularized in the Contract with America that the Republican leadership in Congress set forth to "change the nation" for the better in 1994. Heinzerling demonstrates that the tables showing the costs of risk regulation were "a modern urban legend, such as stories about rats served as hamburger." A large number of the costly regulations on which the estimates were based never took effect, and the estimates by opponents of regulation of the costs per life saved tend to be as much as a thousand times higher than those of regulatory agencies. Further, the high estimates deflate the probable numbers of lives saved and greatly discount the costs over time. Finally, the benefits of regulation include not only saving lives, but preventing "many human illnesses that cannot be quantified . . . [preventing] ecological harm . . . [and preventing] harms to values that are widely shared such as autonomy, community and equity."[93] In 2003, the Office of Management and Budget under the Bush administration reported that regulation is in fact cost-effective.

Without doubt, regulation has costs that must be borne by businesses and citizens alike. Governments must avoid " "technologically forcing legislation" (legislation requiring polluters to use the most advanced technology available to deal with a problem) merely because the technology exists. And they must avoid focusing energy and resources on environmental and safety problems where it is unclear what the outcome will be. Still, pluralist regimes have largely demonstrated their ability to weigh the costs and benefits of environmental legislation, to ensure greater equity in spreading the costs and benefits among all members of society, and to protect the environment—the wetlands, the forests, the plains, and the rivers—from further destruction, so that future generations of citizens can enjoy the natural world.

The general sense is that, in pluralist regimes at the beginning of the twenty-first century, governments and their citizens have made substantial progress toward slowing environmental degradation and understanding its sources. In the mid-twentieth century, streams and rivers were polluted, air in many locales was literally black with smoke, and dangerous wastes had found their way into the soil and groundwater, in some cases because of the intentional practices of disreputable businesses. By 1970, however, government officials had begun to address these problems successfully throughout North America and Western Europe. Laws on clean air and clean water provided substantial fines for pollution and required cleanup. Government officials welcomed procedures to enable ordinary citizens, scientists and engineers, and businesspeople to gain access to the policy process. Officials accepted the view that promoters of technology had to demonstrate its safety and efficacy, and that the environmental costs of production could be managed properly. The archetypal factory of the early years of the industrial revolution, one that billowed smoke and released hazardous waste, had been replaced by means of production that provided jobs without destroying the environment for future generations. Projects by such organizations as the Army Corps of Engineers that sought to transform nature on a massive scale no longer gained rubber-stamp approval, for a variety of reasons.

A change had taken place in our worldview, a growing understanding that all human beings are within nature, not outside it, and that it is a risky business indeed to attempt to improve upon nature. The battles continued between those who promoted nuclear power to satisfy growing energy demand and those who feared its potential for catastrophic failure; between those who worried that governments had established too many rules and regulations that infringed upon individual rights and the ability of businesses to operate properly and those who recognized that the "invisible hand" of the market cannot protect the environment without the assistance of laws and regulations; between those who praised the establishment of concession stands in national parks and permission for snowmobilers to explore wilderness flora and fauna and those who sought quiet and fresh air; and between those who reveled in the comfort of a modern lifestyle and those who criticized its basis in rampant consumption. In pluralist regimes the battles have been resolved in the public sphere, in the courts, and not in backrooms, to the benefit of a few.

But policy makers, engineers, scientists, and other citizens must continue to press the battle to preserve the environment. They must think about how sustainable their throwaway patterns of consumption are and must address the question directly, when the public good might take precedence over some notion of private property rights. And those people must ask whether there can ever be technological solutions to resource use and pollution patterns that are technological in origin. As Rachel Carson reminded us, "The 'con-

trol of nature' is a phrase conceived in arrogance, born of the Nean-
derthal age of biology and philosophy, when it was supposed that
nature exists for the convenience of man."[94] Surely, democratic regimes, with their ability to involve citizens in open discussion of solutions to the ongoing environmental crisis, are equipped to abandon such arrogance toward nature.

CHAPTER TWO

THE COERCIVE APPEAL TO ORDER: AUTHORITARIAN APPROACHES TO RESOURCE MANAGEMENT

How has the role of the state differed in such authoritarian regimes as National Socialist Germany, Communist China, the former USSR, and Brazil under military dictatorship, with regard to management of natural resources? Are authoritarian regimes more effective than democratic regimes in allocating those resources among competing claimants? Are nondemocratic governments able to take full advantage of scientific expertise to move ahead on development projects, absent needless and uninformed meddling from the public? Can such regimes, having set aside concerns about private property, use land in the name of the greater social good to promote large-scale projects—to advance transportation, power generation, and communications technologies? Can these regimes achieve goals of modernization—almost always meaning rapid industrialization—in a shorter time and with fewer of the costs that plagued England, the United States, and other capitalist nations (squalor, dangerous working conditions, profligate exploitation of resources, cutthroat competition)? If historians and political scientists long pondered such questions as these, now they ask, What explains the extensive environmental degradation that seems to be the rule in authoritarian regimes? Is China on a path toward destruction of nature that will exceed the irreversible costs of development which characterized the former Soviet Union?

In this chapter I discuss the way authoritarian regimes have pursued aggressive resource development programs that, without exception, have had significant environmental and human costs and few checks on their momentum. I will focus on large-scale geoengineering projects to alter the face of the earth and its rivers, including hydroelectric and irrigation projects, and the development of extractive industries. The cases of China and the USSR are important to consider because of their leaders' embrace of Marxist ideology and centrally planned economies to secure national goals. I will describe the special obstacles to resource development and environmental preservation that such nations with large peasant populations and numbers of indigenous people as Brazil (and the former Soviet Union and China) encounter. The case of Nazi Germany is unique because of the conflict between technological rationality and the appeals to unreason that characterized the state and its approach to nature and treatment of non-Aryans. It also was a modern industrial power, unlike the others considered here. Yet by any definition it was an authoritarian regime whose policies had a great impact on the environment, and on the millions of innocent victims it slaughtered or enslaved.

By authoritarian regimes I mean those in which, generally speaking, a single political party controls important institutions, including the media, often with marked intrusion into the private lives of individuals; positions in the party and positions in the government overlap; the state maintains a secret police force; and the government is closely involved in the planning of economic activities that benefit it. Authoritarian regimes may be based on institutions of private property (as was true, for example, in Brazil or Nazi Germany) or on state ownership of the means of production (as in the USSR and China). Most of these regimes—although of course not Germany—were at an earlier stage of industrial development than the United States, Britain, or France when they embarked on modernization.

Authoritarian regimes have often claimed to serve the citizenry where other regimes failed to do so. For example, Marxist governments represented the interests of the working class, and development programs were supposed to reflect those interests, by overcoming poverty and providing housing in ways that market economies could not. These nations often were confronted with difficult decisions about how to allocate resources—for example, between heavy industry and public health—that often left the latter shortchanged. For many of these regimes, the struggle to fulfill citizens' needs and the simultaneous necessity to rebuild from war and civil war put great strain on the economy.

One major question about authoritarian regimes, therefore, is whether public involvement in decision making about economic development plans is the only way to ensure both measured economic growth and environmental protection. That is, if political authorities systematically exclude the public from decisions about what resources to develop, how quickly to develop them, and which people or groups of people should benefit, and if the authorities declare that the resources belong to all citizens or must be developed according to some manifest destiny, is it likely that the environmental and social costs of development will be higher than in open political systems? Another question is whether the economic, social, and environmental costs of development in authoritarian regimes have been significantly higher than in other countries. Or have they rather been able to compress those costs into a shorter time frame, taking advantage of the experience of the industrialized nations that preceded them? That is, industrialization (and the attendant problems of urban filth, pollution, and uneven development) took forty years in the USSR but seventy years in the United States and

Great Britain. (See Table 1 for a schematic listing of the foregoing discussion.)

Are authoritarian regimes less sensitive than democratic regimes to indigenous peoples? We know of the costs to Native Americans in the United States and Aborigines in Australia of conquest and settlement. The settlers displaced the Indians, stole their land, and introduced smallpox and other diseases that decimated their populations. In authoritarian regimes, too, scientists and government officials worked in concert to tame resources in faraway regions (for the USSR, Siberia; for Brazil, the Amazon) for the benefit of citizens and the state. But these citizens usually lived in urban centers—

Table 1 Authoritarian Regimes and the Environment

Worldview
• One official view
• State-directed development that determines "progress" for the masses
• A group with special status (for example, workers or Aryans)
• Nature as real, knowable, and important to the state
• Capricious nature as inimical; rational nature as planned

Economic Imperatives
• Large-scale, centralized projects
• Planners' preferences
• State ownership, considered most efficient and just
• Assumption that private property (citizens') is less important than state property

Political Desiderata
• One-party regime
• Secret police with great power
• Dangers to regime, both domestic and international
• Control of media, which citizens have few rights to question
• Regulation tilted to the benefit of the regulated
• Weak, proxy, or nonexistent nongovernmental organizations

Science and Technology
• Technological style that emphasizes mitigation over prevention
• Fewer redundancies (backup systems) in designs for safety
• Restricted input from experts into determinations on safety and efficacy
• Technology as panacea and symbol of modernity

Moscow, Rio de Janeiro, and Beijing—while the costs of development were borne by inhabitants of the periphery. Those inhabitants include the Khanty and Chukchi of Siberia and the Parakana and Yanomami of Amazonia. Further, although development programs are based on extensive scientific study, in authoritarian systems the studies carry the cultural and political baggage of highly centralized management systems and are based on a belief in the infallibility of scientific knowledge that results in an engineering ethos of "victory over nature" at all costs. Indigenous and local peoples have been pushed aside, their histories and cultures irreparably altered. But progress that glorifies the state, as manifested in Siberian rivers diversion project or the Three Gorges hydroelectric power station in China, go ahead full speed.

THE USSR: "HERO PROJECTS" FROM STALIN TO BREZHNEV

In the effort to modernize as rapidly as possible, Soviet leaders put great stress on the development of natural resources and the productive capacity to transform those resources into goods and services important to the state. The emphasis was placed on increased power production and on the mining, metallurgical, and construction industries. The reasons for rapid modernization included fear of what was termed hostile capitalist encirclement. Even if no war was at hand, Soviet leaders believed that war was inevitable, and this viewpoint required the creation of an industrial and military powerhouse to fend off attackers. Soviet leaders therefore placed their hopes on three types of industry: pig iron, steel, and the nonferrous metals; coal and oil to fire the blast furnaces and boilers and to run the engines; and construction and transport.

Under Joseph Stalin (1879–1953), leader of the USSR from 1929 until his death, the single-minded pursuit of heavy industry limited investment in the agricultural, health care, and consumer sectors.

This emphasis also meant that little attention was given to conservation. Officials ordered resources to be exploited as rapidly as possible. Trees were felled for fuel, ore was extracted, and fish were hauled in. The problem of waste was rampant in all sectors of the economy. The state also considered workers and peasants expendable. As a consequence, injury and fatality rates mounted at all construction sites and factories. Housing was very poor; laborers lived in shacks and hovels. By the eve of World War II, policy makers considered natural resources limitless, if difficult to develop. At the same time, officials viewed capricious nature and climate as "enemies of the people," for standing in the way of human plans.

There were several reasons that Soviet leaders adopted a breakneck approach to industrialization. In addition to fear of attack by an advanced capitalist nation, these included the Marxist economic imperative to build modern factories to free men from heavy physical labor and seek economies of scale that had been absent in backward cottage industry. The Soviets believed that technology was the highest form of culture and recognized that technological achievements would serve them well in their propaganda battle with the West. The insular, conspiratorial party apparatus also guaranteed that they would find enemies everywhere and would tolerate no slacking.

A central focus of Soviet development projects from the time of Vladimir Lenin (1870–1924), the leader of the Russian Revolution, was so-called hero projects. The projects included the State Plan for Electrification (1918, known in Russian by its acronym GOERLO), the Dnieper hydroelectric power station, the Moscow metro, the steel city of Magnitogorsk, and dozens of other large-scale technological undertakings. They served both technological and ideological functions. In addition to contributing to the tasks of resource extraction, transport, or manufacture, the hero projects served as forums to educate peasants and workers about the glories of the Soviet state; as sites to transform poorly educated individuals into well-trained employees; and as objects of national pride: these projects were better, the leaders claimed, than any in the West.

Because of the large-scale, state-sponsored approach to resource management, Soviet leaders rejected conservation. They rejected it in part because they believed that the proponents of conservation were largely "remnants" of the tsarist regime, and therefore potential "wreckers." A vital ecology movement had existed in the Soviet Union before Stalin's rise to power. Leading biologists had established a series of professional organizations whose goals included the creation of inviolable nature preserves *(zapovedniki)* to maintain diverse ecosystems and advance the study of them. Under Stalin, party officials came to view these specialists in many cases as dangerous to the state, since state plans for development collided directly with the environmentalists' agenda of preservation. Yet perhaps because their goals for preservation were rather limited or perhaps because officials viewed them largely as innocuous, the state did not subject the environmentalists to the fate suffered by many other people in the USSR: internment in the infamous Gulag, or labor camp system, as enemies of the people. As Douglas Weiner has documented, the environmentalists managed to hold their community together and make significant contributions to ecology during the seventy years of Soviet rule.[1] Still, planners had their way with nature.

How did the environment stand in the way of plans? River flow varies by season. In the USSR the heaviest flows occurred during the spring thaw and the autumn rains. But Stalin demanded that water be available year-round. The nation therefore embarked on an ambitious program to transform the major rivers in the USSR—the Volga, Don, the Dnieper, the Kama, and the Svir—into a unified system of dams, reservoirs, irrigation systems, and hydroelectric power stations connected by canals. The reservoirs stored water so that it would be available for agricultural, industrial, power generation and transport purposes. Huge construction organizations moved from site to site transforming rivers into machines. At these so-called hero projects—at the new steel mills of Magnitogorsk, on the White Sea–Baltic Canal, and at such as others as subways and skyscrapers—party ideologues exhorted young, untested workers to dig, pour, and build faster and faster, not to give nature—its rivers, forests, fields, or ore deposits—a moment's rest. The workers too rarely rested. Many of them died from exposure to the elements or from industrial accidents, especially the slave laborers employed in the construction and mining brigades of the far north and Siberia. Soviet leaders also sought to transform fisheries and forests into industries in short order, by harnessing scientific studies to economic plans.

The culmination of Stalinist designs on the natural environment was the promulgation in October 1948 of the Stalinist Plan for the Transformation of Nature. The plan envisaged the natural environment of the European USSR rebuilt into a machine. Canals, locks, dams, and hydropower stations changed the physical regime of all major rivers. The Communist leadership authorized the planting of tens of thousands of kilometers of "forest defense belts"—trees planted to block winds, retain soil moisture, and assist in the transformation of the southern steppe region into an agricultural machine. And the rebuilding of industry from its destruction during World War II was carried out at top speed. Pollution control, filters, proper disposal of wastes, including radioactive wastes—all these were minor concerns.

Simultaneously with the transformation of the Volga River basin, engineers in the United States had developed the Columbia and Tennessee River basins in similar large-scale projects. Both the Soviet and the American projects resulted in dislocation of local populations and histories, destruction of falls and rapids, and inundation of hundreds of thousands of hectares of land. Once the American projects were completed, the workers were left to find jobs for themselves, whereas the USSR guaranteed employment in massive construction organizations that almost autonomously moved on to other large geoengineering projects. Owing to their scale and momentum, Soviet projects resulted in far a greater assault on the environment and in higher injury and fatality rates among the workers than did similar projects in the West.

THE KHRUSHCHEV ERA AND LAKE BAIKAL

By the mid-1950s, under Stalin's successor Nikita Khrushchev (1894–1971, and leader from 1955 to 1964), virtually all the European rivers had been tamed, to power the Soviet economic machine. They had also become dead zones of pollution and lost their anadromous fish, which could not climb past the dams. Khrushchev did not so much abandon Stalin's costly hero projects as embark on his own. And since the European USSR had been largely tamed, Khrushchev turned the attention of planners toward Siberia. This made sense also from a strategic view. Siberian resources cried out to be exploited, and Siberia would require industrial and agricultural development in the event of another invasion such as that in World War II when German armies had destroyed the European USSR.

A key reason for the extensive environmental damage that characterized the USSR manifested itself at this time. The large-scale construction trusts that had been established in the Stalin era had grown ever larger. The trusts extended to all areas of economic activity: subway and highway construction, the extractive industries, hydroelectric power stations, and more. The trusts had grown from five thousand to fifty thousand and even eighty thousand employees. The trusts built towns for the workers and their families, along with apartments, schools, and stores. In this centrally planned economy, no unemployment was permitted. No sooner was one project finished than planners had to designate another major project to commence, to keep the workers employed. The trusts gained substantial momentum, moving through the countryside in search of the next geoengineering project. The only way a trust could shrink in size or scope of activity was when a new trust was formed from it. The directors of the new trust lured young workers to the first construction site with promises of apartments and schools to be built—but those workers gained jobs in the most difficult conditions and where housing lagged. If you look on a map of the USSR you can follow the construction trusts up the rivers, as they built dams and locks, year by year and decade by decade. In a market system, this kind of momentum is lacking. Once a project is complete, the workers become redundant, and they themselves must relocate to find employment. This may be a check on rampant, rapacious development in market economies in which policy makers, entrepreneurs, and bankers must secure project approval even before hiring construction crews.

The Khrushchev era is known for two major areas of activity that had a significant environmental impact. The first concerned the effort to rejuvenate agriculture. Production had slowed since the collectivization of the peasants in the early 1930s. Although the collective farms were intended to be more efficient and modern than the small peasant plots that had preceded them and although they became a model for American agribusinesses with modern machinery on farmland that stretched to the horizon, the Soviet experiment with collective farms failed. Peasants hated them, refused to work with any initiative, and even slaughtered half their livestock on being ordered to join the collectives in the early 1930s. The farms were also an instrument of political control over the peasants. So despised were the farms—and therefore so ineffective— that at least two famines took a toll under Stalin. A series of droughts and crop failures in the mid-1950s led Khrushchev to institute several crash programs. First came the decision to mechanize agriculture and to use more chemical fertilizers and biocides. The chemicalization of agriculture resulted in the use of up to three times more chemical input per hectare than in the United States, leading to extensive pollution in runoff and to accelerated erosion. Another initiative was the "Virgin Lands" campaign, the decision to plow up the grasslands on the steppes of northern Kazakhstan and western Siberia to grow wheat, a move that resulted in short-term increases in production. But given the insufficient farmers or equipment to keep the lands producing, and the harsh chemical methods used on the steppe, significant soil erosion again resulted. Granted, similar problems often arose in Western agricultural systems, but there chemical use tended to be a third to a fifth that in the Soviet Union, and agricultural extension services allowed Western farmers to be well informed about how to minimize erosion and which crops to plant.

The second area was the transformation of the Siberian landscape, especially its lakes and rivers. Soviet planners, seeing inexhaustible forest, ore, and water resources east of the Urals, built several of the largest hydroelectric power stations in the world on the Ob, Angara, Irtysh, and Amur Rivers. This construction had immediate consequences for the local environment: the inundation of thousands of square kilometers of land. The stations powered the burgeoning metallurgical and aluminum industries. Ore removal was a crude process, which entailed no reclamation of land. Smelting occurred in inefficient, smoky, outdated furnaces that had poorly operating filters. Forests were clear-cut close to existing roads and railroads. The most notorious projects of this sort involved the construction of paper mills on the shore of Lake Baikal near Irkutsk and a plan to divert the flow of Siberian rivers into Soviet Central Asia and the European USSR.

Lake Baikal was a major battleground for nascent environmentalism in the USSR. Lake Baikal, seventy kilometers southeast of Irkutsk, is by volume the largest freshwater lake in the world. It has 2,500 species, 1,500 of them endemic, including freshwater seals. Its one outflow is the Angara River. In Russian literature and culture,

Baikal is the jewel of Siberia. When in 1954 military and paper-industry officials announced plans to build a pulp mill on Baikal's shores to produce a fiber with applications in aviation, poets, writers, and scientists joined to prevent the project. They and other citizens wrote letters to government officials. Articles in literary journals and newspapers attacked the plans as short-sighted and morally and ethically inappropriate; the cultural "thaw" of the Khrushchev era permitted this kind of discussion, which had been impossible under Stalin. The opponents of Baikal development succeeded in getting the government to appoint several commissions to look into the matter, but each commission came back with a recommendation to move ahead, albeit with better regulations, standards, and equipment, to ensure compliance with environmental laws. The paper mill was built, and it polluted Baikal heavily.[2] It and several other mills on the Selenga River, which flows into Lake Baikal, are scheduled to close, but this has not occurred. The battle over Baikal indicated that other battles might be joined, but that the state, ignoring public opinion, which it shaped and controlled almost exclusively, would have its way in most matters. In all, Soviet environment and quality of life for the Soviet citizen declined in the 1950s and 1960s.

THE BREZHNEV ERA AND DIVERSION OF SIBERIAN RIVERS

Pollution accelerated with economic growth in the Brezhnev era. Leonid Ilich Brezhnev (1906–1982) saw the USSR gain military parity with the United States during his years of rule, 1964 to 1982. He also pursued Siberian development programs, including river diversion and the conquest of the Baikal region and the taiga, which he jump-started through the construction of a new Trans-Siberian railroad to the Pacific Ocean. He also presided over increasingly rapid deterioration of air, land, and water quality. The impact of industrial pollution on Soviet citizens led one specialist to liken the epidemiology of the USSR to "ecocide."[3] On paper, Soviet environmental achievements were significant. They included the strengthening of legal codes, the extension of protected lands, and the promulgation of citizens' rights to bring legal action against scofflaws and to live in a clean environment. But in reality it was always cheaper for polluters to pay fines and continue polluting than to control pollution or improve their equipment, and few suits had any consequence for them. The production targets fixed in Moscow were paramount.

By the end of the Brezhnev period the flow of many rivers had slowed to a crawl. Industrial and municipal pollution made them unfit for humans and animals alike. Fish populations had been destroyed. Pollution from heavy metals filled the atmosphere. A British diplomat once told me that because of the automobile exhaust and industrial pollution, living inside the ring road around Moscow, within sight of the Kremlin, was the equivalent of smoking two packs of nonfiltered cigarettes daily. Officials overlooked the pollution problems because, they declared, they were building an industrial superpower for the benefit of the masses, economic growth *would* benefit the masses, and the glorious Soviet constitution protected the environment.

One of the most astounding efforts to rectify man-made problems with another man-made technology was the infamous Siberian river diversion project. This project epitomized the Brezhnev-era commitment to rebuilding nature to reflect state-established goals. By the early 1960s, planners and policy makers recognized that a significant water shortage existed in the European USSR and in Soviet Central Asia. For industry to expand in the former and for agriculture, especially cotton and fruit, to expand in the latter, planners needed to secure more water. The rivers west of the Ural Mountains had already been used to capacity. Engineers and scientists in dozens of ministries and research institutes came up with the audacious plan to transfer 5 to 12 percent of the flow of some

Siberian rivers through massive canals to meet demand in Central Asia and the European USSR.

The Siberian rivers—the Ob, the Irtysh, the Angara, the Enesei, and others—flow from south to north into the Arctic Ocean. Planners and scientists determined that the water was being wasted—it flowed to the sea without being used. Over the course of two decades some 250 different organizations and tens of thousands of engineers and other employees went to work on designing transfer canals, up to 2,500 kilometers long, to bring water westward. To build the canals, they planned to use more than earth-moving equipment. They considered employing scores of small nuclear devices, or peaceful nuclear explosions (PNEs), to excavate earth and rock. (The U.S. Atomic Energy Commission and the Army Corps of Engineers in the early 1960s explored using PNEs to build the Tennessee-Tombigbee Waterway through much more densely populated Mississippi and Alabama).[4] The Siberian river diversion project would obviously have caused significant local and global environmental damage, by reducing river flows, transferring flora and microbes through the canals to new regions, and creating the usual mess of massive construction projects.[5]

Again, as in the case of Baikal, writers, scientists, and nationalists joined forces to fight the diversion. Among them was Sergei Zalygin, a former hydrologist who was by the 1980s the editor of the liberal journal *Novyi mir* (New World). He wrote blistering attacks against the engineers and their project. Commission after commission continued to approve the diversion project, until Mikhail Gorbachev became general secretary of the Communist Party. He finally put the massive project in mothballs, although to this day some Russian officials speak wistfully about it.

In addition to the vast quantities of pollution from such metallurgical centers as Nikel, Norilsk, Magnitogorsk, and Sverdlovsk, the coal regions of the Donbas and Kuzbas regions, and municipalities and industries that discharged untreated wastes into rivers and holding ponds, one of the worst environmental offenders in the USSR was its huge nuclear enterprise. Just as in the United States, engineers promised that they would find a permanent storage facility for the millions of cubic meters of liquid and thousands of tons of solid low- and high-level radioactive waste. Instead they poured the waste into temporarily holding ponds which leaked or leached waste into the groundwater. They jettisoned reactor vessels in the ocean. In a forty-square-kilometer area near the confluence of the Techa and Misheliak Rivers containing two hundred storage sites, they stored roughly five hundred thousand tons of solid wastes and up to twenty thousand cubic meters of liquid wastes. Between 1949 and 1956, highly radioactive waste entered the watershed at the source of the Techa. In September 1957 a nuclear waste dump at Kyshtym exploded, sending millions of curies of concentrated military radioactive waste into the atmosphere, requiring the evacuation of eleven thousand people, and creating a dead zone of several hundred square kilometers.[6] In the hot, dry summer of 1967, when Lake Karachai evaporated, winds blew the radioactive dust more than fifty kilometers, and forty-one thousand people were affected.[7]

High levels of radioactivity exist in many sites of the nuclear enterprise—its power stations, fuel fabrication facilities, and military bases—and must be tolerated by personnel who live and work nearby. But many cities far removed from nuclear establishments also have levels of radioactivity that exceed the norm. Dumps and storage facilities for low- and high-level wastes were poorly constructed and poorly monitored. In many cases they lacked even rudimentary fences to keep out passersby.[8]

The Chernobyl disaster, perhaps more than any other event, indicated the environmental bankruptcy of Soviet nuclear policy—and of its environmental policy generally. The USSR suffered from a significant imbalance between energy resources and population. Seventy percent of the population and industry were concentrated in the European part of the country, and roughly the same percentage of hydropower, coal, and other resources lay in Siberia. Such leading specialists as the president of the Soviet Academy of Sciences, Anatoly Alexandrov, pushed a nuclear solution to the imbalance problem. The nuclear lobby was powerful because of the role of nuclear industry in securing military parity with the United States and allowing the USSR to compete with the United States for prestige in the international arena. Alexandrov and others suggested building a network of nuclear reactors near major population and industrial centers. The policy led to the establishment of reactor "parks" of four to six reactors sharing various basic equipment, located near densely populated areas. It had at least two other questionable consequences, as well: the premature effort to mass-produce pressurized-water reactors; and the design of reactors without containment vessels that might limit the spread of radioactivity in the event of an accident. Engineers also designed and built breeder reactors to generate plutonium and built nuclear steam facilities, located within city limits, to heat apartment buildings and provide energy for industry.[9]

Alexandrov was the inventor of the Chernobyl-type "channel graphite reactor," known by its Russian acronym RBMK. The advantages of the RBMK—relative ease of construction in units up to 2,400 MW, on-line refueling, and the production of plutonium that might be used for military purposes—were significantly outweighed by the absence of containment and the reactors' inherent instability at low power. But Soviet leaders determined to build RBMKs, dozens of them. To this day, four operate outside St. Petersburg.

The Chernobyl plant, eighty kilometers north of Kiev, had four reactors, and the intention was to construct ten in all. As the result of a poorly thought-out and still more poorly conducted experiment in April 1986, unit four exploded, catching on fire and sending a radioactive plume into the air that spread throughout the northern hemisphere. An area thirty kilometers in diameter around

Chernobyl had to be evacuated, perhaps permanently. Dozens of people were killed instantly. Farmland was ruined. Specialists estimate that between five thousand and fifty thousand people will die from cancers as a result of Chernobyl.

The Chernobyl accident revealed serious shortcomings not only in the nuclear industry but also in the Soviet approach toward resource development and Soviet technological style generally. Engineers came to believe that technology was perfectible, that accidents were the responsibility of workers, and that designs ought to focus on mitigation, not prevention, of accidents. They viewed calls for greater precaution in the engineering process as an attempt to resist inevitable progress. And they believed, like their Western counterparts, that nature should be engineered.

THE SOVIET MODEL IN EASTERN EUROPE

The East European countries that fell under Soviet rule through military occupation largely embraced the Soviet technological style, in part because they were coerced into it. The Soviet investment model that favored heavy industry over light industry and collectivization of agriculture to take advantage of mechanization and chemicals was adopted virtually without exception. Thus, once again the metallurgical and mining industries took precedence over the supply of consumer goods and public health issues. Housing construction and reconstruction from war damage lagged considerably. Overcrowding was the rule, and sewer and plumbing systems were often an afterthought. Rivers were dammed in spite of the efforts of those in fisheries industries to protect them. When national Communist leaders collectivized agriculture, they alluded to the goal of making it more modern and efficient. Yet heavy reliance on mechanization, chemicals, and biocides led to destruction of soil and watersheds. As for industry, since rapid industrialization was the goal, the authorities provided investment for increased capacity and production, but little for pollution-control equipment.

The result was coal, petroleum, chemical, and iron industries from Poland to Bulgaria and from Romania to Czechoslovakia that spewed smoke and particulate matter over the globe.

The Soviet state-building model also involved the embrace of nuclear power based on Russian models: VVER (the Russian version of the pressurized water reactor, or PWR) models in Bulgaria, Germany, and Hungary, and RBMKs in Lithuania. Consider the example of the Kozloduy nuclear power station on the shores of the Bulgarian Danube, with its two 440-MW and two 1,000-MW PWRs. Construction commenced in 1974 and coincided with a push by Bulgarian Communist Party leader Todor Zhivkov for Bulgaria to embrace the "scientific technical revolution" of its Soviet comrades. Over the next decade, a propaganda effort to stress the modernity of Bulgarian citizens filled the country's streets and buildings. Posters appeared linking young Bulgarian men and women with images of nuclear power, space, and computer technology. After the breakup of the Soviet Union, officials decided to keep Kozloduy open, even though concerns had grown that the reactors were unsafe. Bureaucrats criticized reactors in Europe and the United States as being less safe than Bulgarian ones and insisted that electricity supplied by nuclear plants was the cheapest and safest available.[10] The Paks Nuclear Power Station in Hungary, also equipped with VVER reactors, remains open, but German safety equipment has been installed. These former East-bloc countries were ill prepared to deal with long-term storage of radioactive waste.

As Robert Darst has shown, the opportunity to sell nuclear technology in the West led Soviet engineers to improve the safety of their pressurized-water reactors, to promise better control and monitoring equipment, and to build containment vessels in second- and third-generation reactors. The Soviets outbid Western reactor firms to build two reactors in Loviisa, Finland. The Soviets offered to sell VVER reactors with extremely favorable financing, with Finnish participation at each stage of the project, unlike with a turnkey plant, and with the Soviets being responsible for disposal of spent fuel. The Finns insisted on full containment and a state-of-the-art emergency core-cooling system.[11]

The story was different in Ignalina, Lithuania, where the two largest Chernobyl-type reactors in the world, each operating at 1,500 MW, were built. Although they ran satisfactorily through the early 1990s, they now operate at half power, to ensure a margin of safety against an accident. Storage of spent fuel rods has become a significant problem. Also, the European Union has made shutting down the Ignalina station a condition for admitting Lithuania to the E.U. Its officials assert that the RBMK is inherently unsafe and fear the accumulation of ever more radioactive waste. Lithuania worries that closure of the reactors will leave it dependent on other countries for expensive oil imports. Further, tens of thousands of workers would lose their jobs.

The nuclear power stations were important sites of environmental protest activity in the period leading up to the breakup of Soviet Union. Jane Dawson explores the way many of the individuals in Lithuania, Ukraine, and elsewhere who first agitated for independence in the Gorbachev era (1985–1991) seized upon environmental issues—in particular, nuclear power stations and other technological manifestations of Soviet power—as symbols of Russian economic exploitation of its empire. The activists believed that Russia had based the most environmentally costly technological facilities and processes in the satellite nations, to keep Russia better off. When the USSR broke up, these environmental cum independence movements lost their momentum because a drop off in state construction projects and the economic downturn directed attention away from ecology.[12] Environmental movements have lost strength throughout Eastern Europe and the former Soviet Union since the early 1990s. The weak economy has left citizens more concerned about jobs than about the environment. The environmental situation has nevertheless improved in many cases, however, largely because that very downturn has led to the closing of many factories, including the most heavily polluting. Whether government regulations will require rebuilt factories to install pollution-control equipment remains to be seen. It seems evident that the countries of eastern Central Europe—Poland, Bulgaria, Romania, and the Czech Republic—have been able to modernize their industries rapidly to compete in European markets. Some have become members of the European Union.[13]

NAZISM AND PROTECTION OF NATURE

Around the same time that Stalin assumed power based on one-party rule in the USSR, Adolf Hitler began his rise to power in Germany, establishing the horrific Third Reich (1933–1945). Hitler also relied on a single party, the National Socialists, or Nazis, and secret police to establish his murderous government, which was based on racist principles about the superiority of the German, or Aryan, people. The Nazis proposed to transform nature, including humankind. They first turned on their own people, through a series of sterilization laws (modeled on U.S. statutes) aimed at ridding the nation of the "unfit." Next, the National Socialists enacted statutes on euthanasia, to eliminate those people—often handicapped, disabled, or retarded—whose lives they considered "not worth living" because of the burden they placed on society. Through the so-called Final Solution, the genocide of six million Jews—and the extermination of many others, including Gypsies, Communists, and homosexuals—the Nazis attempted to cleanse nature of yet others deemed subhuman.

Aside from the devastation the Nazi armies wrought on the nations and peoples of Europe and Africa, what did authoritarian Nazi rule mean for nature preservation and environmental protection? The Nazis' racist and nationalistic worldview had far-reaching effects. According to Raymond Dominick, National Socialism permitted reactionary and racist elements to co-opt nature protection movements. The National Socialists appealed to the organic *völkisch* aspect of conservation campaigns, that is, by claiming that nature conservation maintained the health and welfare of the *Volk* (the people); the Volk was the idealized and idolized German peasantry, the *Blut und Boden* (blood and soil) of the regime. Dominick observes a strong ideological overlap between nature conservation and National Socialism in the country's regeneration after the debacle of the Versailles treaty—the regeneration, that is, of the German Volk and of the organic link between the Volk (blood) and nature (soil). Under National Socialism antimodernist, antimaterialist, and anti-technological elements and appeals to anticommunism thrived, all of which gave support to several environmental movements. As Jeffrey Herf discusses, German scholars, engineers, and political thinkers held complex ideas about the place of technology in German culture. On the one hand, Germany was arguably the most technologically advanced nation in the world. Its chemical industry had afforded economic self-sufficiency, even in the absence of natural resources, in a number of strategic cases. The National Socialists relied on automobiles and highways to promote economic growth, and used science and technology to achieve unrivaled military power. And yet many German thinkers rejected modern technology as somehow un-German, in part because of their belief that industrialization had somehow disrupted the organic "blood ties" between the Volk and the land.[14]

Granted, scientific ecology was foreign to the Nazi creed, and Hitler's militarism was alien to nature conservation. Yet völkisch ideology found a favorable breeding ground at the intersection of nature, race, and militarism. The Nazis believed that the German people, especially the peasants, had "defined their character in interaction with the soil of the Fatherland." Nature was the source of vigor for the national soul and the foundation of the people's character. The Nazis and nature protectionists who embraced the

Greater-German outlook called for all Germans—for example, those in the Sudetenland in Czechoslovakia, which Hitler annexed in 1937—to be repatriated and for the land to be returned to its true owners. As for racism, some conservationists believed that the air pollution that damaged Germany's forests and the water pollution that killed its fish had been produced by the "unwashed brood of Croats and Pollacks" who had displaced the blue-eyed Aryans. Hitler went far beyond these conservationists when he called for nature to be conquered "naturally" in the name of the Volk.[15]

What happened to the conservationists among the Nazis? In spite of the emptiness of the Nazi position on conservation, most conservationists joined forces with them. As Dominick explains, they saw close connections between love of nature and love of the Fatherland. When the Nazis attempted to co-opt conservation clubs—in the same way they strove to control or subjugate most other social and political clubs and organizations—the results were mixed because of infighting. As a result, Nazi officials succeeded in penetrating some societies only slightly, and those societies lacked the typical Nazi regimentation. But conservation movements—and individuals—on the political Left, largely unlike those on the Right, faced intimidation, violence, even death.[16]

Ultimately, as Dominick shows, conservationists seem to have avoided membership in right-wing parties, although many joined after the Nazis came to power—indeed, at a higher rate than adult males generally, or than medical doctors, teachers, or lawyers. Among German professionals of all kinds, foresters had the second-highest rate of participation in Hitler's party, trailing only veterinarians. Still, only one item was "shared *uniquely* by Nazism and Nature conservation: the conviction that Nature shapes the national character." Dominick concludes, "Only one kind of conservation, the völkisch variety of *Naturschutz* [nature protection], was centrally and durably aligned with Nazism."[17]

When the National Socialist regime took up nature conservation, part of the reason may have been to avoid the deficiencies in enforcement and protection characteristic of the Weimar Republic. Another reason—on paper—was to make nature conservation more accessible to the general public. Yet the Reich Nature Conservation Law (1935) shifted responsibility for conservation from the regional governments to the national government, which acquired veto power at that time over decisions taken at lower levels of government. This power enabled the Reich government to harness nature (and all strategic resources) for industrial and agricultural purposes. Only nominally did conservation remain in force.[18]

Surprisingly, in its first years the Nazi government supported several important conservation measures. They included rejection of a proposal for a railroad right-of-way in the Bode Valley, a protected region since the 1890s, and publication of an ordinance on animal and plant protection that had languished since the 1920s. German leaders gave symbolic support to other conservation efforts on both the local and the national level, including cooperation on the establishment of new nature preserves. Still, the Imperial Conservation Law of 1935 gave Hermann Göring and the Ministry of Justice control over conservation matters. This weakened the Office of Nature Protection considerably. At the same time, the government undertook massive public works projects, including the Autobahn (highway) system to jump start economic recovery and promote autarky. As Thomas Zeller has explained, the highways were supposed to be aesthetically consonant with the environment. Needless felling of trees was to be avoided and the highways were to have gentle curves, to follow the landscape.[19] Yet as the labor force for the Autobahn grew to one million people, earth moving became a major occupation, and the number of automobiles also grew rapidly—all developments contributing to irreversible environmental damage. Still, nature reverence persisted in the ideology, even as the Nazi government embarked on a destructive world war.[20]

Another aspect of the National Socialist worldview, the concept of *Lebensraum* (living space), had significant ecological implications. Lebensraum was crucial underpinning for the Nazi war machine. According to Hitler and his followers, it was German destiny to conquer the lands to the east—and the racially inferior Slavs, Jews, Gypsies, and others living in them—so that the Third Reich could grow. The Germans would then exploit the conquered lands for their agricultural products and forestry. The conquered peoples would serve as slave laborers for them. The natural environment in Germany proper would remain organically sound, if not pristine, for extraction of resources would occur elsewhere. The government set up research offices to plan the exploitation of the land conquered in pursuit of Lebensraum.[21] In National Socialist Germany, then, the institution of private property did little to protect the environment. Planners within the government saw nature in all its manifestations as something to feed the military machine (the Wehrmacht).

It may be, therefore, that belief in the sanctity of private property is not the crucial ingredient in preventing environmental devastation, even if individual property owners strive to protect their land and seek legal or extralegal redress from those who threaten it. Rather, state power is the key—that is, how well it fosters protection of private and public property through laws and regulations and how well it gives access by the public to legal proceedings. Further, if resource development is state-sponsored and state-directed, grave dangers remain that such development will be misguided, will be carried out in the name of one group at the expense of others, and will be large in scale, as the case of China indicates.

DOES THE CHINESE DEVELOPMENT MODEL DIFFER FROM THE SOVIET?

Perhaps because China's hydrologists were trained at Soviet institutes, Chinese engineers pursued costly nature transformation projects with the same vigor as their teachers. A centrally planned economy where planners' preferences prevail and a one-party political system that excludes the public from participation in the technology assessment process have also contributed to the adoption of megaprojects that are destroying the Chinese environment. Whether old-style Communist leaders under Mao Tse-tung or the new generation of Communists who welcome capitalist reforms while holding firm in the battle against political liberalization, Chinese officials have always believed that economic growth is a sine qua non; consequently, all obstacles to it—social movements, regulatory statutes, environmental quality—must be controlled, overturned, or ignored. The extraction of resources in the countryside for the benefit of city dwellers, the poverty of peasants, who are increasingly required to fend for themselves, and the absence of a coherent policy to balance resource development for the benefit of all citizens exacerbate the degradation of nature. According to some measures, China now pollutes more heavily per capita than any other country in the world.

According to Judith Shapiro, Communist Party chairman Mao Tse-tung's approach to nature was similar to Stalin's, in that he saw struggle against the earth to be "an endless joy." Maoist efforts to transform society and the economy resulted in environmental disaster—for example, through misguided reclamation efforts, extensive deforestation, and the construction of poorly designed dams. In the absence of opposition from engineers (let alone from the largely illiterate and powerless peasantry) irrational projects moved forward. Indeed, as in the USSR, the supreme ruler and the Communist Party, distrusting expert knowledge, forced many specialists into the countryside (in China, during the Cultural Revolution) to toil as peasants, and imprisoned or executed others. The Great Leap Forward of the late 1950s was an example of this irrational and costly approach to China's modernization under Mao, as production of steel in small-scale blast furnaces spread pollution far and wide and accelerated deforestation.[22]

Prohibited by the Communist Party from discussing the great environmental and social costs of Mao Tse-tung's economic strategy until after his death, specialists began only in 1979 to publish in a series of technical and political forums evidence of the extensive degradation that had occurred since the Communist takeover in 1949. On the surface, all indices reflected superb achievements. The battle against wasting water was particularly successful. If in 1949 only a few reservoirs exceeded a hundred million cubic meters, by the mid-1980s three hundred such impoundments and more than fifteen hundred medium-size ones were available to store river water for power, irrigation, and general water supply. They were, however, poorly managed. They filled up with silt more quickly than had been estimated, which shortened their operating lifetimes considerably.[23]

China also had the advantage of more than fifty thousand rivers with basins larger than a hundred square kilometers. (Much of the arid interior has no permanent rivers and no drainage to the ocean.) The Yellow and the Yangtze (now Chang) Rivers are the most important, both overdeveloped for irrigation, hydroelectricity, and municipal purposes, like the Volga River in the USSR. Owing to this overdevelopment, and to a mismatch between agricultural activity and technological use of the river basin, Chinese hydrological engineers have long discussed transferring water from the Yangtze to the Yellow, the first such grandiose scheme having been set forth in the late 1950s under Soviet influence. Engineers discussed three major projects, each of which would involve building a canal roughly a thousand kilometers in length, which would permit the irrigation of millions of hectares. The Grand Canal was completed in 1990, but it transferred only one-fifteenth of the amount of water anticipated by the most enthusiastic of the planners.[24] Barely a third of Chinese agriculture depends on irrigation, as opposed to half or more in North American and European countries. The paucity of irrigation projects results from the Chinese failure to plan and carry out projects carefully to the end.

In the Three Gates (Sanmenxia) project on the Yellow River, engineering overconfidence and miscalculation combined to scuttle ambitious plans. Egged on by Soviet hydrologists, the Chinese approved a plan in 1955 to build forty-six dams on the river that would impound water to produce 110 billion kilowatt-hours (kWh) annually. The Sanmenxia facility was 110 meters high and 839 meters long, creating a 3,500-square-kilometer reservoir. The hydroelectric plant was rated at 1,100 MW. Planners, anticipating problems with silting, built thousands of silt check dams and silt precipitation dams to prolong the plant's life by fifty to seventy years. Yet nearly 60 percent of the dam silted up within just fifteen years. Power generation halted, but even removal of the turbogenerators and abandonment of the water storage facility did not help. Today Sanmenxia operates at just 250 MW and is idle much of the time.[25]

Overly ambitious plans and the exaggerated expectation of engineers that they could conquer any unforeseen problem at Sanmenxia should have been a lesson for the ongoing Three Gorges Dam project on the Yangtze River. China has a centuries-long history of water impoundment projects. Since 1949 the state has resettled roughly ten million people to make way for dams and irrigation networks. In 1949 there were about forty small hydroelectric power stations in the country; by 1985 the central government had forced the construction of eighty thousand reservoirs and seventy thousand stations. Three of them alone—Sanmenxia, Danjiangkou, and Xin'anjiang—compelled a total of one million people to move. In most cases conflict arose between the residents forced to move and the authorities because the government never provided sufficient resources to relocate or retrain those ousted or to build adequate housing, schools, or hospitals for them. Usually people were moved to land that was less productive.[26] We shall see that Three Gorges was possible only in an authoritarian system.

Technocrats promoted the Three Gorges Dam for several reasons. One is flood control, since major floods have affected the Yangtze River basin every ten years on average. Those of 1931, 1935, and 1954 killed 320,000 persons in all. A second is improved river transport to permit thousand-ton boats to travel between the east coast and industrial Chongqing Province. Of course electrical energy production is the third and main reason. With a capacity of 18 to 22 GW (gigawatts), the hydroelectric power station will help China bridge a gap in supply and capacity. Electrical energy production in China is 5 percent that in the United States and 20 percent of Korea's.[27]

The Three Gorges Dam illustrates the dangers of state-sponsored industrial and agricultural development when people are viewed as factors no different from capital. Proposed initially by the nationalist leader Sun Yat-sen in 1912, the dam grew in the final plans to nearly two thousand meters long with a reservoir stretching six hundred kilometers behind it. Relocation has been slow; the new home sites are hovels; and the compensation for displacement is inadequate. As of 1989, the office for poverty relief of the Ministry of Agriculture acknowledged that roughly 70 percent of oustees lived in "extreme poverty," and 46 percent had yet to be properly resettled. Diversion of the Yangtze River started in November 1997; the reservoir began to fill late in 2002, and it is projected to be full by 2009. Officials have announced the need to relocate 1.2 million people before the reservoir fills. Two large cities (Wanxian and Fuling), 11 country towns, and 114 townships will be submerged. Many experts predict that in fact up to 1.6 million individuals will have to move, and that claims that 200,000 persons have already been resettled are inflated. Further, the decision to build the dam is questionable from the point of view of loss of farmland. On the surface, there is good justification. China lags behind in availability per capita of fresh water (2,484 cubic meters, by contrast with the world average of 7,744 cubic meters) and averages only .08 hectares of arable land per person, as opposed to .26 hectares per person worldwide.[28]

For the peasants, the future could not be bleaker. They lack basic information about where they are moving and what they will face, or how much compensation they will receive. Indeed, is it possible to compensate individuals financially or psychically for the loss of homes and burial grounds? In some counties, local officials have embezzled resettlement funds or taken bribes from the builders responsible for roads, schools, apartments, and reclamation of new farmland. The dam will inundate 632 square kilometers, including more than 300 square kilometers of farmland, and not the rich rice-growing terraces usually associated with Yangtze Valley, but steep hillsides dependent on rain. Resettlement will require the ousted residents to go to still higher ground, where the steeper terrain and poorer soils are prone to erosion and inadequate for agriculture. Since that new land is less fertile, the resettlers will require perhaps five times as much land as previously, or they may have to buy expensive farming equipment. The government planned to find employment for peasants in local industries. But no industry has been built, so unemployment is rapidly increasing; only token unemployment compensation is offered.[29] All this environmental devastation and social dislocation has taken place in the name of progress.

Promoters of the Three Gorges Dam earned degrees in Moscow at the prestigious Zhuk Hydroelectric Design Institute. The Soviets named the institute after the infamous Sergei Zhuk, who began his career by using forced prison labor to build canals and other failed monuments to Stalin, for example the White Sea–Baltic Canal (Belomor), where tens of thousand died in arctic conditions using hand tools to construct a failed technology. Was this progress? For their part, the Chinese engineers pushed the Three Gorges project from the 1950s on. Three decades later such central Chinese bureaucracies as the Science and Technology Commission, the State Council, the Ministry of Water Conservancy and Electric Power (a bureaucracy whose essential conflicts of interests exist even in its name), and such international organizations as the World Bank had completed feasibility studies in support of the project. Their feasibility studies indicated the benefits of electric power generation, improved navigation, and reduced downstream flooding yet all but ignored the costs of construction: more than 630 square kilometers will be inundated, and this is the most fertile and useful land in the region. Flood control may be a major goal, but the dam will hold back twenty-two billion cubic meters of water, which is only half that loosed in a disastrous flood in 1954. There will be large increases in downstream sediment load (an increase in turbidity) to bring about lower sedimentation above the dam. The delta sediment will become finer, creating fluidized mud. The dam may produce sixty billion kW/h annually, but the planners have calculated neither the cost of lines, transformers, and other equipment nor the demand load. And since electrical energy cannot be stored, it cannot be used efficiently. Only during one or two months of the year will the elevation of the headwaters be sufficient to generate maximum turbine power. Finally, the reservoir will take seventeen years to reach full capacity in any event.[30] No one seems to have asked, Are there better ways to generate electricity and use land?

Because of the generally accepted view that large-scale projects are desirable, little discussion took place in China or in the West about the potential dangers of Three Gorges. Between 1984 and 1989, when Chinese officials and engineers were debating the Three Gorges Dam, and when government agencies in the United States and Canada, seeing lucrative opportunities, actively sought contracts or offered to provide feasibility studies, American media coverage of the dam was "limited . . . and superficial." In China, the coverage was no better. *People's Daily* itself carried only six articles between 1985 and 1990 related to the project, all in support of it.[31] The Chinese government employs four full-time journalists to convey the state's message that all is in order. Three Gorges is a hero project, no different from those of Stalin, about which criticism is barred, and the citizenry must march in lockstep behind it.[32] Jackson and Sleigh described the Three Gorges Dam as "unfeasible in the 1960s, unaffordable in the 1970s, and politically and technically opposed in the 1980s." They note that the public is much less compliant now than it was in the Maoist 1950s. But after the authorities put a violent end to dissent at Tiananmen Square in 1989, Three Gorges moved to the top of the state development agenda.[33]

Besides the Three Gorges Dam project, China's policy makers have made continuing environmental degradation likely in several ways. These include growing reliance on coal, in conjunction with expanded economic growth; inadequate waste management and recycling programs; and a system where the crucial factor in determining a project's worth is its economic benefit to the leadership. The leaders equate economic benefit to themselves with trickledown benefits in the form of economic growth (jobs in an export economy) and believe that economic growth will bring peasants and workers into the twenty-first century. Post–Mao Tse-tung reform efforts to promote projects included decentralization to eliminate bureaucratic inefficiencies. In the electric industry the results have been mixed. Like the former Soviet Union, China seeks increased electrical energy production, which it sees as central to future economic growth. Unfortunately, because the government embraced too many expansion projects at once, including those for electrification, resources were insufficient. Capital investment in transmission rose but use outstripped generation capacity. The shortfall of resources has shifted the burden to provincial and local governments to generate electricity, and onto small, decentralized, and often inefficient and heavily polluting plants.

In 1982 more than eighty thousand hydroelectric plants with a total capacity of 8,080 MW were in operation in China. China's electrical output is fueled primarily by coal (53.5 percent), with 14.8 percent from oil and 31.6 percent from hydroelectric sources. Coal reserves are estimated at six hundred billion tons, but 40 percent has sulfur content of 2 percent or higher, and 20 percent of reserves has 3 percent or higher. Sulfur dioxide emissions have risen rapidly, decade by decade, and no environmental controls affect most of the coal consumed, whether in factories or in homes. To meet growing demand, hydropower has become a major source of energy in rural regions, with projects under 500 kW on small streams with low head (the difference between the height of the water impoundment and that of the water outflow) given priority. That is to say, every available site may be developed. As for coal-fired thermal plants, the option in other regions, the focus is on 12-MW units adapted from a larger, 100-MW design. But many of the critical components, such as controls, boilers, and water treatment and cooling systems have not been or cannot be similarly engineered to achieve efficiencies commensurate with reduction in capital costs, and the plants lack modern burners or filters. The only advantage is that these units can be constructed quickly.[34] In regions with limited hydro-resources, China operates two thousand of these small, highly polluting thermal power stations.

THE PEOPLE'S FISH ALSO SUFFER

The advantages of a centrally planned economy are alleged to include an end to disputes between officials in different sectors, a check on bottlenecks and inefficiencies in the allocation of resources among them, and a brake on high prices determined by market demand, which might prevent individuals or areas of the economy from acquiring a given product. These advantages do not in fact exist. Planners' preferences ensure that certain forces will win out in the battles between sectors over resources. The environment, consistently undervalued, has been the loser in China.

A good example concerns coastal resources. The conflicts in China between manufacturing, mining, shipping, and fishing industries, and the environment has been won by the polluters. Land reclamation for industry has destroyed fisheries. Since the 1970s, China has made coastal development a high priority, for the simple reason that the country's eleven coastal provinces have 41 percent of the population, if only 13 percent of the nation's land, and contribute 54 percent of the gross domestic product (GDP). Exports from these provinces have grown rapidly since the early 1980s. Because foreign trade and domestic economic growth trump all other considerations, policy positions that might slow growth have no place. Like property rights advocates in the United States, Chinese industrialists believe that these include environmental protection legislation, which should be avoided or ignored. As is typical in authoritarian regimes, each ministry or sector has responsibility for economic development, and environmental legislation—in this case, that affecting fisheries, seaports and harbors, ores, and energy—is legally required to minimize any adverse impact. But ministries do not have adequate funds, personnel, or willpower to do so. Further, no body exists to oversee coordination among sectors or to mediate disputes that arise between them, although the State Oceanic Administration is supposed to ensure "comprehensive ocean management."[35]

Indeed, Chinese planners and managers see the coast as a resource waiting to be tapped. To pursue export markets, China expanded its shipping from 350 million tons in 1986 to 480 millions tons in the 1990s (a 37 percent increase). Shipping required new moorings and therefore encroachment on traditional fishing areas. Nearly 1.2 million hectares of tidal lands disappeared to agriculture, salt-making fields, shrimp ponds, and ship berths between

1950 and 1986. More than 160 placer mines along the coast and in the water also contributed to erosion and pollution, as did twenty major, official offshore dumping sites for hazardous waste and garbage. To make matters worse, in the absence of regulations and fines, annual discharges of industrial and municipal waste reached about 6.5 billion tons annually, including 200,000 tons of pesticides, not to mention dozens of major oil spills. The Yellow, Changjiang, and Zhujiang River estuaries, and the Dalian, Laizhou, Haizhou, and Jiaozhou Bays have been particularly hard hit. Red tide and other harmful algae blooms occur frequently, bringing about a sharp drop in the number and diversity of invertebrates. Shellfish beds and mangrove communities have simply disappeared.[36] In pluralist regimes, such violation of human rights of the disadvantaged people who live along the cost, such flaunting of regulations, and such callous disregard for ecosystems—in a word, such steamrolling of nature, would not be tolerated.

Pollution is the most significant problem that makes it difficult to deal with the many competing demands on limited water resources in China. Daily discharges of polluted water were roughly seventy-eight million cubic meters in 1979; two-thirds of monitored streams were polluted. This country of hundreds of millions of urban residents had only thirty-five small municipal water treatment plants in 1980. More than 90 percent of urban wastewater is discharged untreated. Twenty-five times more polluted water is discharged in Shanghai than can be treated. Sewers often break down, having rarely been prepared properly. People remove night soil by bicycle, handcart, and oxcart, or in buckets on poles. The stench is unbearable. The slop leaks, splashes, and overturns into the street. Few garbage trucks are available to help in the task. Bureaucrats take no interest in sanitation. From peasants in houseboats and from factories, the poisons flow forth: untreated waste, phenol, kerosene, and heavy metals. Funds for monitoring or enforcement are inadequate, for the plan and production are king and queen. Air pollution is no less serious, in part because China is the third-largest consumer of primary energy, behind the United States and the countries of the former USSR (although on a per capita basis, China ranks much further behind). Chinese power plants operate with half the efficiency of U.S. and European facilities, and rarely do they scrub the coal beforehand. Acid rain is a serious and growing phenomenon.[37]

The forests too have fared badly in closed systems, whether the USSR, Brazil, or China. In China this is surprising, because the National Environmental Protection Act of 1979 and the National Forest Act of 1984 led to the involvement of large numbers of scientific specialists—ecologists, geographers, economists, and others—in reforestation programs to ensure sustainable forestry practices. Yet once again, demand for wood for fuel and construction trumped environmental considerations. China has never been rich in forests. In 1976 they comprised only about one-eighth (1.22 million square kilometers) of its land area. Almost half of the Chinese forests lie in the northeast and southwest, and they have been so poorly managed that roughly 70 percent are categorized as "aged," 23.7 percent as "middle-aged," and 7 percent as young (saplings). Because of fire, disease, and careless harvest and transporting practices, survival rates are low. A quarter of the forest is affected by insects and other pests. Soil erosion and floods are endemic problems. As in the former Soviet Union, infrastructure (that is, roads, railroads, and equipment) is decades old. Reforestation programs intended to address this situation have been poorly funded and carried out. A forestry extension system to provide scientific information and modern equipment to forestry trust employees remains rudimentary. And not surprisingly, the government has failed to enforce existing laws.[38]

As in the former Soviet Union, deforestation is the result of cutting that annually exceeds natural growth rates by threefold, of illegal felling in the absence of enforcement, and of phenomenal waste in manufacturing. If paper, pulp, and woodworking plants manage to utilize 50 percent of the tree, that is considered a lot. Highly touted reforestation projects are significantly behind plan. The stripping of vegetation to make way for agriculture and urbanization has accelerated erosion, lowering the quality of life of peasants in particular. Fifteen percent of the nation has suffered serious soil losses, while 11 percent is already desert, and that percentage is growing. The result is almost a Chinese dust bowl.[39]

All of this deforestation has accelerated the extinction of even common species. Land under conservation constitutes less than 0.2 percent of all Chinese territory, and little has been added over the past decade. Few Chinese are educated in ecology; no recognized university curriculum existed in the environmental sciences before 1979. Even though public awareness of environmental problems is growing and citizens have begun writing letters to officials and newspapers to protest the situation, little gets done in the rush to produce necessities for the growing number of Chinese citizens and to meet demand in burgeoning foreign markets.[40] China may have achieved the dubious distinction of being the most environmentally devastated nation in the world. Air and water pollution top the list of problems. Outdated boilers, smelters, and power plants spew particulates—sulfur, nitrogen, and carbon dioxides—into the air. Rivers, lakes, and coastal regions have increasingly become dead zones. Forests and other resources are harvested without regard for the future.

THE MARCOS REGIME IN THE PHILIPPINES

Other authoritarian regimes have turned to large-scale development programs with great political, social, and environmental consequences, not for the benefit of the citizenry, in spite of rhetoric to the contrary, but for the state leaders. Under Ferdinand Marcos (1917–1989), Filipinos experienced fifteen years of authoritarian rule and impoverishment in the name of a struggle against Communist insurgents and Muslim terrorists in the Philippine Islands.

Efforts in the 1930s at land reform through the National Land Settlement Administration (NLSA), to encourage "colonization of the sparsely populated areas in the Archipelago," predated the Marcos regime. The NLSA had as its goal to hasten settlement of the hinterlands. With its large budget, through the Rural Progress Administration it promoted land tenure among peasants, to protect them from unjust eviction.[41] World War II interrupted the NLSA efforts, and Marcos's predecessors had little further success in following through on promises to empower the peasantry through land reform.

Marcos became president of the Philippines in 1965, having ridden to victory on promises to improve living conditions for Filipinos, in part through land reform to provide peasants with land tenure and ownership. He pushed a major public works program. But deterioration in the quality of life in the Philippines in the late 1960s led to public demonstrations, violence, and insurgencies. Student demonstrators joined the opposition. This reaction led Marcos first to suspend habeas corpus and then to declare martial law in September 1972. Relying on coercive measures and the support of the military and secret police, Marcos then succeeded in changing the constitution to have himself declared president for life. His wife, Imelda Marcos, became governor of Manila and minister of human settlements and ecology. She was known more for her extensive collection of shoes than for her achievements as minister. The government ordered the arrest of opposition figures, among them journalists, students, and labor activists, closed down newspapers, and otherwise controlled the mass media.

While Marcos and his cronies spoke about the way in which a free-market economy based on land reforms would create democratic institutions, they instead used martial law to reduce freedoms and nationalized manufacturing enterprises or gave them to

Marcos associates and relatives who stole the profits. The army and secret police nearly trebled in size and used their powers to intimidate, arrest, torture, and sentence. The accused had little legal recourse. Perhaps sixty thousand people were arrested between 1972 and 1977. In 1986, after another rigged election, public outrage became so great that the Marcoses left the Philippines for the United States and a democratic regime took over.

Through an ideology of progress, modernization, and anticommunism, Marcos advanced thousands of state-supported projects to promote industrialization and modernization of agriculture. The Development Bank of the Philippines financed farming, cottage industries, and other projects to "hasten the development of the rural areas." Marcos believed that development was "the greatest weapon against subversion." The Marcos regime fostered the National Power Program as part of its effort to overcome "dilemma of the third world"—illiteracy, poverty, and susceptibility to insurgencies. The program included a number of projects for plants with centralized, large-scale generating capacity, with transmission lines radiating out from them.[42] Rural electrification and irrigation required the development of river basins, and as a result the displacement of peasants and tribes. For hydroelectricity, the goal was the introduction of 4,210-MW capacity in Mindanao and isolated areas of northern Luzon, while coal-fired electricity would grow from 55 MW to 260 MW by 1987. The Coal Development Act of 1976 provided incentives to producers through subsidies and loans to abandon pick-and-shovel mining for modern large-scale excavation.

As a sign of his will to modernize, Marcos also pursued nuclear power. The failure of nuclear power indicates that other development projects were more appropriate for the Philippine economy than nuclear power. In July 1973 the Marcos regime decided to build six nuclear power plants. One reactor was planned for the Bataan Peninsula with a loan from the U.S. Export-Import Bank. Taking advantage of the corrupt regime, Westinghouse Electric

Company used a Marcos crony as a lobbyist to outplay General Electric in the bidding process. Westinghouse proposed to build two 620-MW reactors for $650 million. Ultimately, even with little construction completed, cost overruns and interest charges had increased the cost to over $1 billion for one reactor. The site selected, one of high seismic activity, only five miles from a volcano, was abandoned after Marcos's flight to America. The plant may eventually find use as a fossil fuel facility.[43]

In most regards, state-directed development and ownership by Marcos and his cronies followed upon martial law. Marcos established a National Economic and Development Authority to formulate and oversee economic development. A National Grains Authority replaced the Rice and Corn Administration. In agriculture, the Marcos regime turned to green revolution rice and other crops to promote self-sufficiency supported by "massive credit, technological assistance programs and a fertilizer price subsidy." The green revolution would "break the cycle of poverty." Marcos welcomed investment from the Asian Development Bank, the World Bank, and the International Development Association. His regime pursued large-scale cotton and sugar production to generate export income. The attendant social and environmental disruptions were the same as elsewhere with these crops. The industrial sector (mining, quarrying, manufacturing, construction, and utilities) topped all other sectors of the economy in investment and growth. Other efforts included mining of gold, silver, copper, chromium, manganese, nickel, and mercury. Modern communications technologies, as well as public works for transport (roads and bridges) appeared, though based on insufficient planning, to open links "between urban areas and the hinterlands."[44]

One case, pond aquaculture, illustrates the dangers to local inhabitants and the ecosystems in which they live when authoritarian regimes promote development in the name of progress, democracy, and economic growth but in fact support principally the enrichment of the leaders and their coterie of economic allies. Pond aqua-culture, which has existed for centuries in Southeast Asia, became an export-generating technology in the Philippines under Marcos. In the 1960s and 1970s the Marcos government encouraged com-mercialization of pond aquaculture with a view to spreading in-tensive production techniques to provide food for growing urban populations and international markets. This national development strategy had the endorsement of international organizations and banks (the Asian Development Bank and the World Bank), which saw an opportunity to ensure the prosperity of capitalist institu-tions and thereby deny Marxist rebels any foothold. As part of Marcos's administration founded on martial law, the government published a decree in 1972 to encourage the conversion of fields to aquaculture. Industry was the major beneficiary; lenders, conserva-tion officers, and enforcement agencies assisted them. Those who endorsed this "blue revolution" claimed that it would create em-ployment, increase food production, and generate export earnings. They also justified it as improving on nature, which they considered to be wasteful and inefficient. The technology of mangrove fisheries operates in brackish water fishponds that produce prawns, milkfish, and tilapia, usually in a monoculture. The ponds acquire fingerlings or larvae, raised in small nursery ponds that are then transferred to larger rearing ponds; fish are harvested four to six months later. In the Philippines the sector grew rapidly to $1 billion in 1992—that is, grew more than corn, coconut, and other agricultural goods. Having decided to transform nature, and facing opposition only from poor coastal fishermen, engineers cleared intertidal mangrove forests to build ponds.[45] Had the fishing communities and environ-mentalists been involved in discussions about aquaculture, they would have rejected the Marcos government's plans.

Only after deposing Marcos and creating a democratic govern-ment were officials and bankers able to address the costs of aqua-culture openly. Local communities had suffered along with the mangroves. Approximately 46 percent of the country's original mangrove swamps had been converted to ponds; mangrove forest cover has declined fourfold since 1920s. Philip Kelly's research on the Bataan catchment basin reveals that authoritarian aquaculture destroyed a complex local lifestyle and ecosystem in the name of national progress. The fauna—crabs, prawns, and other crustaceans—that were staples of the local diet and trade were lost. The destruction of mangroves led local residents to turn to coconut for firewood; it was highly polluting. Further, the ponds generated little local employment. They require few caretakers and do not foster local ownership. Since pond yield is sold to capitals, local food supplies have become more expensive and less varied.[46]

Seeking increased production at any cost, authoritarian regimes tend to ignore the social impact and environmental degradation caused by modern fishing technologies. Yet fishing technology—whether it is deep-sea trawling using sonar to locate, kilometer-long drift nets to catch, and floating factories to process the harvest from the deep or more traditional methods of catching coastally based salmon, trout, shrimp, and eel aquacultures—has grown so powerful that all regimes must learn to regulate fisheries carefully. The farmed fish do not notice whether they are grown under authoritarian or pluralist regimes, but the fact that they require significant expenditure of energy supplies to grow and substantial reworking of ecosystems to thrive means that governments must join with business interests to create a legal and financial framework in which they can operate.

Large-scale approaches to resource management problems and to economic development were the choices of leaders of authoritarian regimes. These approaches were based on the notions that nature can and ought to be transformed into a machine that operates according to plan, that few limitations affect human designs on nature, and that government officials and their scientists know what is best to do. The governments were unresponsive to protests about

their choices—indeed chose to conduct close surveillance of nascent environmental NGOs and excluded them from participation
in the policy process.

AUTHORITARIAN REGIMES AND RESOURCE MANAGEMENT IN SOUTH AMERICA

For much of the twentieth century, Brazil was an authoritarian regime, ruled by representatives of agricultural and industrial elites
or, in the 1960s and 1970s, by the military. Beyond its rain forest,
Brazil has extensive reserves of iron, aluminum, gemstones, and
other resources. But land is most important to the urban elites, who
see its value in a source of lumber for the cities or cleared pastures
for raising beef cattle. The rivers running through the forests are
valued for their hydroelectric potential. European conquerors and
settlers first arrived in the sixteenth century, but in the absence of a
strong state, it remained for their twentieth-century descendants to
apply modern science and engineering prowess to the task of taming the Amazonian interior. A series of rubber booms and busts in
the late nineteenth and mid-twentieth centuries that triggered the
construction of some road and railroad infrastructure was the first
step. After World War II the government turned in earnest to Amazonian development. Government officials employed the standard
methods of other authoritarian regimes: large-scale projects approved by the central government with little public input; five-year
plans intended to secure rationality; and construction of roads, hydroelectric power stations, and extractive industries to serve as a
base for colonization.

As has been well documented, the inundation of settlers accelerated deforestation and the marginalization of indigenous peoples.
When European colonization of Brazil began in the sixteenth century, there were 5,000,000 to 6,000,000 indigenous people; at the
end of the twentieth century only 220,000 remained, and eighty-
seven distinct Indian groups had been exterminated. By 1990 some

twenty-seven million hectares of land had been set aside to preserve the remaining indigenous population, almost half of it since the mid-1980s. Is this too large a price to pay for human diversity, given the amounts of land and resources the government has given to businesses or made available to them at subsidized rates? As in the United States, where mining and forestry interests have fought to maintain privileged access to the federal lands on which they make their profits, so large agricultural and mining interests in Brazil have fought any extension of protected areas.

After the World War II, President Getulio Vargas proposed to integrate the Amazon into the larger Brazilian economy, by creating the Superintendency for the Valorization of the Amazon (whose Brazilian acronym was SPVEA) in 1953. The SPVEA achieved limited results, the most important being the enlargement of the legal extent of the Amazon, the conversion of the wartime rubber bank into a regional development bank, and the first major postwar technology to enter the Amazon, the Belem-Brasilia highway (1956–1960).

Just as Leonid Brezhnev had had designs on Siberian resources to support the lifestyle of urbanized European elites, so in Brazil Amazonian development centered on taming the rain forest and peoples for the benefit of powerful city dwellers. In their respective efforts to fight capitalism and communism, both the USSR and Brazil treated nature as an enemy. Urban residents, especially in Rio de Janeiro and São Paulo, sought capybara, jaguar, iguana, boa, anaconda, and other pelts; and above all else, they needed electricity.

In 1964, after a military coup, Brazilian generals took over SPVEA, purged its membership, and restructured it as the Amazonian Development Superintendency (SUDAM, in its Brazilian acronym) to pursue military and technological control of the Amazon.[47] As in National Socialist Germany, a philosophy similar to the ideology of Lebensraum motivated military leaders to open the interior lands to civilization.[48] The programs SUDAM instituted were based on the notion of the need to bring European civilization, modern technology, and racial superiority to the wilderness, tame the frontier, and integrate "it forever into our national structure."[49] With the same aims, SUDAM aggressively pursued agribusiness and infrastructure in the form of highways, roads, power lines, mining, and logging. The superintendency provided tax incentives to stimulate investment. Its leaders sought private and international investment through such large-scale projects as Operation Amazonia and the National Integration Plan (in Portuguese, PIN).[50] The USSR had used programs such as Virgin Lands and Big Volga to focus national attention and support for the leaders' aspirations, and authoritarian Brazil made use of similar campaigns. The costs to the poor inhabitants and nature itself were similarly extensive, too.

The PIN was intended to "bring men without land to land without men." The "men without land" would enter the interior along fifteen thousand kilometers of newly constructed roads and highways: the four-thousand-kilometer Transamazon Highway through the states of Para and Amazonas to Bolivia, the Northern Perimeter Highway into the northeast, and the Cuiaba–Porto Velho Highway in the southern interior across the states of Rondonia and Mato Grosso. The highways of the PIN, which were poorly constructed, failed to ease the movement of commodities to and from urban-industrial centers. The few feeder roads disappeared at right angles from the highways into mud. The growth of highways accelerated deforestation. Things were no better for the settlements and towns, because central planners ignored local climatic and geophysical conditions in designing the towns. The Amazon ranches themselves are large-scale technologies, covering millions of hectares, and bringing high-technology inputs (for instance, chemicals, fertilizers, and pesticides) to bear on fragile Amazonian ecosystems. The peasants and settlers and squatters reaped inadequate wages, expropriation of their land, and displacement.[51]

The highways brought miners after gold and corporations after aluminum and iron. Whether from small-scale placer mines and river dredging or large-scale aluminum smelters, the result was the same: toxic chemicals from processing flowed onto the forest floor and into the streams and rivers. Government leaders and businesspeople welcomed the highways and did not worry about pollution, because the highways brought "civilization" to the Indians and facilitated the conquest of natural resources. Planners believed that the Amazon was so large that it could tolerate pollution, forest clearing, cattle ranching and other forms of encroachment. Given that mining was a significant aspect of foreign trade and that reserves of gold and other rare metals were estimated at $50 billion, many Brazilians expected encroachment onto Indian lands and pollution to spread even more.[52]

As in the USSR and China, in Brazil planners and policy makers saw big projects as a key to economic development. In 1980, the government established the Grand Carajas program to bring valuable metals and gems—aluminum, tin, copper, tungsten, uranium, iron, gold, and diamonds—to the surface in enormous quantities. Grand Carajas reflected a shift in government strategy for Amazonian occupation from one of encouraging small-scale agricultural settlements to one of stimulating export-oriented growth "poles" based on mining and ambitious cattle-raising schemes. Grand Carajas covered 895,000 square kilometers, an area larger than France and representing more than 10 percent of the total territory of Brazil. In the absence of organized public opposition to their projects, government officials kept economic goals in the forefront, pushing indigenous people aside and subjecting the rain forest to devastation.

The most persistent attack on Amazonia came through national electrification programs. Brazil's leaders saw electrification as the means for bringing civilization to and achieving control over the interior. Hydroelectricity was the choice, and the models of extensive development along the Columbia River basin in the United

States, the Volga Basin in the USSR, Egypt and the Nile, and China and the Yangtze River postwar dam programs were hard to ignore. The Amazon had an estimated one-sixth of the world's freshwater, making hydroelectric power the apparent panacea for development. Engineers identified more than a hundred potential sites for major facilities with potentially scores of gigawatts, the challenges being how to develop them, given the relatively flat relief of the Amazon basin, and how to bring engineers and builders to the sites. (Specialists at the Hudson Institute, a think tank in the United States whose employees believe that free-market economic growth is the solution to all problems, proposed damming the Amazon to produce an inland sea that would cover an area the size of France, and facilitate access to the region's resources.) By 1980, total Brazilian hydroelectric capacity was 25,000 MW, but engineers advanced "realistic" projections of 150,000 MW, with stations costing tens of billions of dollars and claiming hundreds of thousands of acres of land. Soon the engineers sent electricity out along tens of thousands of kilometers of power lines to the cities along the coast.[53] I refer to power lines, highways, railroads and other such technologies that made possible the conquest of the interior corridors of modernization.

The Tucuruí hydropower station, built in the 1970s to fuel the Grand Carajas program and the aluminum plants of the coast, was the first large dam built in a tropical rain forest. To prepare the site, the authorities liberally applied such defoliants as Agent Orange and pentachlorophenol. The dam was built in spite of environmental warnings that turbines would be damaged by the chemical decomposition of inundated forest, that fisheries would be disrupted, wildlife lost, and Brazil nut trees destroyed. Not only water but workers flooded the area, along with roads, tractors, transmission towers, and clearings.[54] Behind the dam spreads a reservoir of some 2,875 square kilometers with a shoreline of 5,000 kilometers. Eletronorte, an electric holding company, owns 6,300 kilometers of towers and transmission lines, which crisscross the forest. The dam, the reservoir, the excavation, the grading and piling, the roads, trucks, barges, bulldozers, dump trucks, and mixers scarred, inundated, unsettled, and destroyed far more of the landscape and did far more harm to the Parakana Indians than the scientists had originally estimated. Tucuruí is an area of hundreds of species of fish, birds, monkeys, armadillos, anteaters, sloths, pacas, snakes, and frogs, all of which are now at risk. This outcome was to be expected from the start, for thirty-eight thousand hectares of Parakana land alone disappeared under the reservoir's waters.

The evils of paternalistic, authoritarian regimes working closely with private entrepreneurs to steal land and resources from indigenous peoples assumed epidemic proportions in the 1960s, 1970s, and 1980s in South America. For example, the Pai indigenous peasant farmers live in Amambay in eastern Paraguay and across the border in the state of Mato Grosso in Brazil. The generals of Paraguay adopted a series of land reform projects and legal statutes to provide the Pai and other indigenous groups with land and rights. They also established the State Indian Body as the bureaucracy to administer indigenous needs and businesses. Yet instead of representing Pai interests, the government looked the other way when contracts permitting the exploitation of Pai lands and people were signed: the contracts allowed indiscriminate logging and gave the Pai few of the proceeds.[55] As in Brazil and the United States, the agencies established to represent the Indians' interests have often instead served the state and big business—for example, by facilitating the giveaway of mineral rights or forestland without payment of adequate royalties.

THE HIGH COST OF RESEARCH DEVELOPMENT IN AUTHORITARIAN REGIMES

While we may debate the similarities of such regimes as the former Soviet Union, the People's Republic of China, and National Socialist Germany, we must observe that their resource and waste management practices have imposed environmental costs. Discussion of these costs should serve as a point of departure for developing nations desiring to become either industrial powers or competitors on world markets for their goods.

In spite of the visible and long-term costs of megaprojects, the megaprojects have yet to go out of vogue. In fact, there has been a renaissance in high dam construction since the 1980s, and not only in China. The renaissance has occurred in nations whose people have yet to organize into cohesive environmental action groups or NGOs, and whose leaders do not intend to tolerate public displeasure about development projects. In Venezuela the Guri GUI project is forecast at 10,000 MW, the equivalent of ten nuclear reactors. In Brazil the Itaipu has a maximum power of 12,600 MW. Hydroelectric technologies are only slightly better understood now than thirty years ago. If anything, the uncertainties about their environmental and social impact have increased with the size of the dams. The experience in each nation indicates that it is nearly impossible to derail a dam whose initial conceptualization may have occurred a century ago but which now has modern state power and vast economic interests behind it.

Especially in authoritarian regimes where public involvement is lacking, the very decision to build stations prevents uncertainties from being fully aired. The experience of the USSR, National Socialist Germany, and China indicates that environment is at best a secondary concern. Hidden beneath ideologies of national destiny, the mission of the proletariat or the Volk, or the need to avoid the evils of capitalism were economic development programs that served heavy industry, agriculture that relied heavily on chemicals, and the military. The interests of the state predominated, even when the ideology spoke of defending the workers' interests or made some other such claim.

In the absence of public involvement in the technology assessment process, projects and strategies in authoritarian regimes reflected planners' preferences. Those preferences made increased production the sine qua non. Social and environmental concerns that might have found their reflection in better housing, support for public transportation, or a slower pace of development were rarely aired, let alone embodied in national plans. The result in Nazi Germany was a war that destroyed much of Europe. In China, Russia, and Brazil, the natural environment—especially the forests and the rivers—has been damaged, perhaps irrevocably. And the public, whose interests development programs were intended to serve, has paid the price, as health and welfare indicators have deteriorated.

The absence of public participation is the crucial difference between authoritarian and democratic regimes. In pluralist regimes, citizens usually have access to privately owned and public media and are able to create or join NGOs to set their own policy agendas and criticize others. Experts who question the efficacy and safety of technologies promoted by the state or powerful economic interests (such as corporations and the military) find a public forum. This openness enables citizens to mobilize effectively to halt, postpone, or require the redesign of projects that they believe are too costly from a social or environmental standpoint, as the examples of disputes over supersonic transport (in the United States in the 1970s), hazardous-waste cleanup, nuclear power, and clean air and clean water legislation demonstrate. Nor have pluralist regimes generally advanced large-scale projects to transform nature in the name of the Volk, the proletariat, or some other privileged group, although such projects as hydroelectric power stations often benefited urban residents at the expense of rural residents and led to the marginalization of indigenous groups.

Such countries as China and Russia that abandoned the strict authoritarianism which characterized their governments in the 1990s have made little progress on developing a worldview more in keeping with conservation or on recognizing the value of public participation in technology assessment. For example, in Russia, the absence of a time-tested civic culture continues to permit economic development programs to run roughshod over the natural environment. The administration of President Vladimir Putin has embraced large-scale approaches to development of resources—for example, oil and gas. President Putin disbanded the Russian environmental protection agency, giving responsibility for monitoring and enforcement to provincial governments, but denying them the funding or personnel to carry through on those tasks. Since then, the Ministry of Natural Resources has returned to such Soviet-style practices as putting development ahead of environmental concerns. The ministry sees it as its mission to open the nation to development and sell off seemingly inexhaustible resources to huge corporations. The Russian government rejects input on policy issues from citizens and from NGOs alike. As a result, nuclear power officials have advanced a crash program to build more than a dozen reactors, without encountering so much as a whimper from the public. Similarly, in China, the economic liberalization of the 1990s and beyond has not been accompanied by limited political openness, and economic growth at break-neck speed trumps environmental concerns. Unfortunately, as I consider in Chapter 3, the absence of access to the policy process or of the media or technical expertise to question or halt projects, along with a legacy of colonialism, has led to widespread environmental degradation in the southern-tier nations of Africa, Asia, and South America.

CHAPTER THREE

DEVELOPMENT, COLONIALISM, AND THE ENVIRONMENT

Southern-tier nations—located largely in Africa, Asia, and South America—tend to be agrarian with traditional patriarchal village and political structures. Under the influence of colonialism, and also by their leaders' choice, these countries often embraced a model for resource development connected with Western Enlightenment notions of science and progress. More insidiously, colonial powers extracted resources from those nations at great environmental cost. The colonizers logged the forests for their shipbuilding, furniture, and fuel industries. Vast tracts of trees were cleared to establish agriculture, which benefited the empire but rarely the peasants forced to labor in the fields. The colonists mined extensively, paying little heed to the environment or the slaves who worked for them. Families were disrupted, the landscape scarred, and the waterways spoiled. The standard practice was for colonial powers to extract raw materials, export them out of the colonies for manufacture at home, and then require the colonies to purchase the reimported manufactured goods.

In this chapter, we shall explore the environmental and social costs of colonialism in southern-tier nations. One question will be, Is it appropriate, inasmuch as colonialism ended in the 1950s or 1960s at the latest, to blame colonial rule for the persistent economic and environmental problems that southern-tier nations face? Surely, population pressures play a role in environmental degradation in developing nations. Those countries which have the highest birthrates in the world tend to have the most limited resources. To support their families, peasant farmers, fishermen, and hunters have put great pressure on those dwindling resources by moving farther into the forests for firewood and into the savanna in search of game. Can technological solutions be found to any of these problems?

Frequently, war and drought have pushed the people of southern-tier nations into a downward spiral of poverty. Warfare accelerates environmental destruction and forces millions to migrate. These refugees then become estranged from familiar lifestyles and ecosystems. The refugees are consigned to squatters' camps that are breeding grounds for fatal illnesses, infections, and disease. Poaching of megafauna has reached epidemic proportions in some areas, where armies of soldiers equipped with advanced military equipment use the proceeds from poaching to purchase more weaponry. War and political instability in Sudan, Chad, Niger, Mali, Mauritania, Mozambique, and Angola in the 1980s led directly to a decline in per capita food production. Countries that had to import food incurred large foreign debts.[1] As Ruffins and Coleman-Adebayo write, these problems stem not from overconsumption, as would be true in the West, but from poverty and war.[2]

In spite of rich biodiversity in the various ecosystems of Africa, nations and the people that inhabit them face an uncertain future because of rapid environmental deterioration, inadequate access to clean water, and inappropriate choices concerning how best to develop their resources. The causes of environmental degradation include the colonial legacy, destructive wars, and in several cases, paradoxically, modern agricultural and industrial technologies that were supposed to be a panacea for the rapid modernization of the continent. To make matters worse, toxic waste, much of it imported from other nations, has been disposed of improperly.

Two aspects of the relationship between environment and technology in postcolonial regimes must be kept in mind. First, in spite of the disruptive impact on colonial systems, technological innovation has had unquestioned benefits. Agricultural technologies have permitted significant increases in and reliability of yields, enabling peasants in many cases to move out of poverty, beyond subsistence levels of production, and into markets where they can sell their surpluses. Rural electrification has been an unquestioned benefit. Nearly everywhere that power lines have appeared—from the hollows of Appalachia under TVA to the villages of Nigeria and Vietnam—they have been accompanied by improvements in public health, longevity, and education. Electricity, as symbolized by the light bulb, has truly been an illuminating force.

Still, electrical energy production also generates environmental change and social disruptions. When thermal power stations produce electricity from highly polluting fossil fuels, and producers are not required by governments to use pollution-control equipment, the toll on public health and the environment are great. The construction of hydropower stations usually leads to the relocation of entire communities and the inundation of fertile floodplains. How can policy makers, businesspeople, engineers, and regulators ensure that the benefits of the dams—flood control, irrigation, and energy production—will be shared by all members of society while other costs are minimized?

No one would recommend a return to the stone ax and wooden plow to lessen the disruption to ecosystems associated with modern techniques and machines: powerful tractors, tillers, and combines. Yet green revolution technologies—here defined as high-yield crops and the irrigation, fertilizer, herbicide, pesticide, and machine technologies required to cultivate them—are capital-intensive, well beyond the means of most peasants, and like most other chemically based technologies must be used cautiously, to guard against soil erosion and ground and water poisoning. And like electrical energy

technologies, green revolution practices usually disrupt village social and economic relationships. To put it simply, international high
technology has overwhelmed indigenous locally based technologies. In many cases, the leaders of African and Asian nations have
determinedly pursued large-scale projects (dams, highways, electrification) that have had a negative impact on nonrenewable resources, on ecosystems (the rain forest, savanna, and fragile taiga
and tundra), and on people (through land stewardship, social structures, health care systems, and so on).

Second, unlike most pluralist and authoritarian regimes, postcolonial states are weak and are based not on transparent but on
hidden political and economic relationships. Their bureaucracies
have little power or influence at any distance beyond their capitals
over policies, and their budgets are inadequate to administer laws
or regulations. Some theorists refer to the system of rule in many
postcolonial regimes as a kleptocracy, where elites rake off the benefits of the economy and demand bribes in order to act on citizens'
behalf. These states were unable to collect long-range data on rainfall, river flow, and so on, necessary to provide a foundation for
development projects—for example, dams and reservoirs to store
river water. In their weakness, contemporary African states resemble their colonial predecessors. Formal colonialism in the Portuguese colonies of Africa continued into the 1970s, and the settler
regimes (Rhodesia/Zimbabwe, Namibia, and South Africa) held on
to power even longer, with political, economic and environmental
instability often the result.

Weak states were vulnerable to nations with advanced technology and extensive capital resources that offered foreign aid, engineering expertise, technological know-how, turnkey plants, or a
combination of these, not only to assist altruistically but to gain influence among political elites and to secure markets for their goods.
The World Bank and other international financial institutions contributed to southern-tier nations' acceptance of large-scale tech-

nological development projects, by failing to understand or by overlooking their significant social and environmental impact. Little attention was given to how smaller investment projects might lessen the negative impact. And multinational corporations (MNCs) took advantage of the political and financial preconditions to enter the new markets with modern technology and offered few alternatives to local people.

All these trends suggest that the environment and the citizen fare poorly when state authority is tenuous. Even as it pursues development, modernization, and urbanization, the weak state lacks adequate policy-making financial, legal, and regulatory institutions to manage the changes that accompany technological innovation. The citizens of many African, Asian, and South American nations thus live in two worlds, one colonial and one postcolonial, with two technologies, indigenous and international, and in two settings, one rural and one urban—with all the attendant social and environmental uncertainties.

AGRICULTURE AND ENVIRONMENTAL CHANGE IN AFRICA

Africa holds some of the world's largest and most diverse ecosystems: vast inland lakes, savannas, dense forests, deserts, and the fauna that live in them. Between the Sahara Desert in the north and the rain forests near the West African coast lie the Sahel and the savanna. "Sahel" is Arabic for "shore." The Sahel is the southern boundary of the Sahara, like the Sahara's seacoast. The Sahel has perennials that provide abundant food, and such fauna as gazelles and desert partridges. A greenbelt of vegetation protects the environment so little wind or water erosion occurs. "Savanna" refers to a treeless or sparsely forested plain. The trade routes that have crossed this region for centuries served several empires that predated colonial rule: ancient Ghana, medieval Mali, and Songhai.

Nomadic and pastoral peoples have raised cattle along the edges of the Sahara for centuries. They and itinerant farmers used hoes and sticks to farm, wells and hand or bucket pumps for water, and portable grinding tools. Low population densities created little need for fences and other boundaries. The inhabitants employed portable mats and frameworks for ease of setting up and packing tents and huts for travel. The success of these agriculturalists in using tools indicates that they "lived by many measures with greater ecological efficiency than modern intensive farming." But African governments, "driven by a desire to increase agricultural production," have encouraged farming of cash crops in marginal areas. This has led to destruction of shrubs, trees, and ultimately entire ecosystems.[3]

Just as in the dust bowl in the United States in the 1930s, meteorological phenomena in Africa have less to do with desertification than do human hands and institutions. In most places, the Sahara does not blow sand into the Sahel, as a result destroying farmland. The sources of desertification are often human—not the farmers or pastoralists—those who lived by herding animals, often lived as nomads and moved according to seasons—but policy makers seeking to modernize agriculture, or elites seeking to control land and people. Careless use—rapacious exploitation in the colonial era and the inappropriate application of green revolution high-yield cash crops and chemical fertilizers in the modern era—has been the major source of rapid change in the ecosystems of the Sahel, accelerating desertification, soil erosion, and deforestation. In Ethiopia alone, forestland has shrunk from 40 percent to 3 percent of the land.

To what extent does the crisis of environment and agriculture owe its genesis to the colonial experience? In what ways has long-term exploitation of peoples and resources for the benefit of European powers during colonialism contributed to the creation of agricultural institutions and approaches that are environmentally unsound? States emerged in West, East, and southern Africa between the tenth and eighteenth centuries. They were small and grounded in religious authority. Iron tools entered agriculture in the twelfth to eighteenth centuries. Cattle raising brought a more settled lifestyle. Agriculture was based on relatively simple technology, by European standards, but was highly adapted to local conditions and was based on centuries of cumulative knowledge about soils, weather, and seeds. When new crops were introduced—corn, bananas, cassava, and peanuts—in the fifteenth and sixteenth centuries, they contributed to security of food supplies moreover complementing millet and sorghum with wild and domesticated fruit, vegetables, and meat.[4]

Some colonial travelers, officers, and scientists reported that the African peasants of the late nineteenth and early twentieth centuries knew their environment quite well. Their knowledge of ecosystems was long-term and comprehensive, helping them succeed in agriculture—for example, through sophisticated intercropping patterns. Other observers were impressed by how much pastoralists understood about grasslands ecology. The peasants recognized soils and their quality by evaluating what grew on them. Their slash-and-burn practices indicated an understanding of fertilization, weed control, crop rotation, and fallow periods.[5] But military conquest and the slave trade overwhelmed traditional knowledge, technologies, and social organization.

Slavery and European exploitation slowed Africa's technological and industrial expansion, hurt agricultural production, and interrupted seasonal farming cycles. Over four centuries Europeans and fellow Africans enslaved roughly twenty-three million Africans. They brought them to slave markets across the Sahara, to the Indian Ocean, and across the Atlantic. Perhaps another ten to twenty-five million persons were killed, injured, or displaced during this time.[6]

Under the pressure of technological change imposed by the European powers, many farmers abandoned traditional agriculture or, having been pushed out of it, increasingly served as migrant laborers, in many cases leaving their families behind. As population grew, the system of leaving some land fallow (the fallow system) disappeared, as did the forests. One result was decline in the per capita output of sub-Saharan agriculture, along with rising dependence on food imports, at the end of the nineteenth century. Yields for such export crops as cotton and peanuts and such irrigated crops as rice, green beans, and maize increased, however.[7]

In the late eighteenth and mid-nineteenth century, European nations began to seek raw materials and markets abroad, including in Africa. During the colonial period (1865–1960) African farmers were drawn steadily into markets that differed significantly from their traditional economies. The colonialists systematically exploited local economies for the benefit of the mother country (through extractive industries, plantations, and so forth). Meanwhile, traditional rural industries—weaving, pottery, and metalworking—were undermined or wiped out. The colonial nations sought raw materials not available at home, and markets for their manufactured goods.

To exploit the African people completely, the colonialists needed to transform traditional, small-scale agriculture into a large-scale venture through plantations. Land became a commodity; cash crops rather than food crops gained in importance. In addition, national and colonial political leaders tended to promote modern agriculture for the benefit of urban elites, at the expense of rural residents, with deleterious effects not only for farmers, but on farmland and forest. For example, cocoa farming in Ghana and Nigeria had relied on family labor. As it became commercialized, enough revenue was available for farmers to hire labor for weeding and harvesting. Modest profits permitted acquisition of land and expansion of farms. Ultimately, only the larger farms that drew strength from their existence as colonial plantations or the co-optation of local elites could hire laborers.[8] Leach and Mearns write, "An overt, social control agenda lay behind policies in East and southern Africa that placed physical restrictions on African farming activities, supposedly in the interests of conservation, because they directly threatened the interests of European settlers."[9]

Another aspect of the negative impact of colonialism on the environment in Africa was the creation of artificial borders. The European colonists imposed bureaucracy on illiterate people who lived in tribes. Most people in sub-Saharan Africa lived within stateless societies—that is, without centralized authority—although they had complex political, economic, and social relationships. Many of them, especially the pastoralists, were nomadic. Great-power relations dictated the boundaries between tribes arbitrarily, often on the basis of the location of their trading towns and military outposts. These boundaries cut across ecosystems, economic and social relationships, and tribal lands, in the process often including mutually hostile groups within the same borders.

Concurrently with the development of agricultural science in the United States and European countries, several colonial powers strove to introduce that science in their possessions. They believed that it would contribute to more rational exploitation of resources. In the settler states of South Africa and Rhodesia and the British colonies, the state intervened in peasant agriculture in the name of science, through forced resettlement and through agricultural departments. Officials were preoccupied with soil erosion and the need to combat it. This concern gave them the justification to impose their conservationist ideas upon the peasants. By the mid-nineteenth century, colonial botanists had already determined "the state of the veld and forests to be so poor as to threaten the future of colonial agriculture as well as many plant species." Scientists called for legislation to conserve the environment. They attempted to promote planting of vegetation where the land was bare from burning, overgrazing, or timber cutting. They drew on information from other parts of the world, especially the United States, whose conservation movement under President Theodore Roosevelt seemed to be devoted to, among many other things, rational farming practices. Also, the dust bowl of the 1930s raised the specter of the lands that might be destroyed by peasant farmers in Africa. Scientists saw African agricultural methods as "careless and dangerous," and they believed that animal husbandry tended toward overstocking because of hoarding of animals for status and dowries. This belief, combined with a perceived shortage of land, led to innumerable commissions, reports, and memoranda that documented the soil problem.[10]

One of the major sources of tension in developing nations, dating to the colonial period, was between the Western scientific knowledge of agriculture and conservation, which was international, and the local knowledge of peasants. Both colonial and postcolonial governments and international development agencies favored the Western view of modernization based on science and engineering expertise that had a significant long-term impact on African people and environment. The governments and agencies tended to ignore locally defined problems and solutions because they did not mesh with national development plans designed by bureaucrats in distant urban centers. National elites that embraced the Western view and their counterparts in international agencies saw peasants as ignorant and lazy.[11] Writing in 1945, Kenyan colonial officials believed that

> the African in Kenya has not yet arrived at the level of education which enables him . . . to plan his agricultural economy successfully . . . In his case, therefore, it is essential that his general farming policy shall, to a large extent, be dictated to him in the light of the experience and knowledge of officers of Government responsible for his welfare.[12]

Clearly, peasants were not lazy, ignorant, or incapable of following government policy. They had accumulated knowledge over centuries of farming in a difficult environment, and they saw no reason to abandon their methods for expensive, unfamiliar ones imposed on them from the outside.

The view of African peasants as an obstacle to modernization persisted through the end of the twentieth century. Many officials who attended the Earth Summit in Rio de Janeiro, Brazil, in 1992 endorsed the view of peasants as the problem and modern technology as the solution when, under Agenda 21, they referred to Africa's farmers, hunters, and herders as both agents and victims of environmental change. The assumption was that if "trends are to be reversed . . . local land-use practices will have to be transformed and made less destructive."[13] That is, they believed that development programs based on expert advice and green revolution technology were needed to change the lifestyle of indigenous peoples. A neo-Malthusian argument (that is, one based on Thomas Malthus's warning in a 1798 treatise that population growth would soon outstrip the ability of society to produce enough food) of a crisis of overgrazing, deforestation, soil erosion, and desertification of drylands, especially under pressure of rapidly growing population, held sway among many African government officials and professionals, personnel of international donor agencies and NGOs, and academics.[14] Leach and Mearns argue that the interests of officials in government, aid agencies, and NGOs and of independent experts "are served by the perpetuation of orthodox views, particularly those regarding the destructive role of local inhabitants." According to crisis narratives, pastoralists and peasant cultivators were a major force of environmental degradation. This view enabled experts and managers to assert their rights as "stakeholders" in the land and to usurp power and resources from local people.[15]

Leach and Mearns identify a series of historical and scientific arguments that undergird the crisis scenario. The typical case identified a shortfall of a resource and its roots in consumption patterns and growth rates and then projected a supply gap into the future (for instance, an alleged firewood crisis in sub-Saharan Africa). Once the crisis was identified, it required government forestry departments with financial and technical support from aid agencies and Western worldviews to intervene to rectify wasteful indigenous practices. Yet, as the case of firewood reveals, the advocates of the "crisis" argument ignored the fact that most wood used as fuel in sub-Saharan Africa is surplus wood gathered from land cleared for agriculture or left after branches were lopped off. And where people do have a crisis, they have always found local ways to respond (planting trees, encouraging natural regeneration of trees, reducing consumption).[16] In essence, local practices and organization of resources have logic and rationality that researchers in bureaucracies far away from the field rarely fathom. This recognition recalls James Scott's argument that states must aggregate knowledge in manageable forms to make policy, but that this aggregation cannot take into consideration variation—including local institutions and social structures.[17] Unfortunately, effective development policies and programs that mobilize funds, institutions, and technologies "depend on a set of more or less naïve, unproven, simplifying and optimistic assumptions about the problem to be addressed."[18]

Yet if the "crisis narrative" has faults, it remains a fact that a significant environmental and social crisis plagues billions of the world's inhabitants, largely in southern-tier nations. At the beginning of the twenty-first century, 70 percent of Africans were destitute or living on the verge of poverty. Per capita income lagged significantly behind that in the rest of the world. Many people were unemployed. Three out of four Africans lacked access to clean water. Between 1961 and 1983 food production per capita declined and consumption stagnated or declined. Grain production per acre dropped. Meanwhile, many developing nations turned to such cash export crops as cotton, while permitting food production to decline.[19] They also adopted Western agricultural techniques, hydroelectric power stations, and other practices and technologies that suggested they remained mired in a colonial-era dependence on Western approaches, science, and technology.

THE CYCLICAL CRISIS OF ENVIRONMENTAL DEGRADATION AND FOOD PRODUCTION

Natural climatic conditions in many regions of Africa contribute to agricultural problems. In some cases there is not enough rainfall, in others too much. Half the continent is arid and desert. There, rainfall is too scarce to support cultivation even of millet. Only 19 percent of African soils are free of "inherent fertility limitations." Extensive soil erosion is a problem for usual reasons (wind, rain, lack of ground cover), and also because the proportion of clay and organic matter in soils is lower in Africa than in most other places. Downbursts of rain in tropical regions often cause erosion. Deforestation facilitates erosion elsewhere. And more than a third of African land is either desert or rocky soil so dry or shallow that "minimum tillage [is] best."[20]

The limitations of resources and environment in Africa cannot be ignored. Take the case of water. Africa has more landlocked countries than any other region of the world, and many African countries have fewer rivers and lakes than do other countries. The situation resembles that in western China, but the Chinese government mobilized its resources and workforce to build canals and irrigation systems. In some African nations, owing to inadequate administrative and scientific planning, growing reliance on crops that require irrigation compounded the problem of inadequate water. Granted, one-third of the agricultural lands are irrigated, and they account for two-thirds of agricultural production. Yet the African continent uses only 4 percent of its available agricultural lands, the lowest percentage of any region in the developing world. Surface water and aquifers are limited, and only the oil-producing countries of the Middle East can afford technology for desalination of ocean water. (Shockingly, Moroccan leaders are considering buying

Russian nuclear desalination technology that would be provided by reactors on barges moored along the Atlantic coast.) Highly efficient trickle, or drip, irrigation appeared in some countries in the 1990s. Water shortages mean that the dates when farmers plow and plant are crucial. In areas with no animals or a low level of mechanization everything must be done by hand, and in many places the men have migrated to the cities in search of work, so women and children must till the fields. As the drylands and local ways of life deteriorate, more and more peasants—especially men—migrate to urban centers, where they create vast squatter communities that have their own environmental problems. Migrant communities often lack sanitation facilities and never have schools or hospitals. The squatters live near pollution-spewing chemical industries. How much more can the women and children who are left behind do to support subsistence agriculture?

Population pressures have been a factor in the disappearance of millions of hectares of forest and woodland each year. Most African nations have high fertility rates and high mortality rates, a combination that fosters dependency, not self-reliance. Africa, both north and south of the Sahara, and the Middle East have annual population growth of 2.8 to 3.0 as opposed to 0.8% annually in the industrialized countries. The continent's population tripled between 1950 and 1990. Compounding the problems inherent in this population growth are some of the lowest literacy rates in the world.[21] Still, India and China have high fertility rates and yet have managed to increase food production to meet demand.[22]

Traditional agriculture maintained ground cover when population densities were low. This practice was more the result of the effort to conserve labor than a conscious decision to conserve soil by leaving in stumps and grasses. A problem is that since independence most nations have undertaken large-scale, high-cost, import-intensive projects that do not involve local populations. The urban-based experts painstakingly designed new fields, then ordered the attack on the ground with bulldozers that destroyed fragile soils. Harrison argues that a low-tech approach would have been cheaper and probably have worked better. It could have included stone lines (to hold and back up water), hedgerows (to retain water and soil and protect against wind), intercropping (the planting of several crops simultaneously in the same field), and mulching.[23]

Would irrigation technology that produced food crops be more appropriate than irrigation employed to develop cash crop markets? And should not the first question be how best to use available workforce and capital resources? Most African irrigation relies on free flooding, which makes use of gravity or controlled flooding, in which water stagnates and seeps in. Both are inefficient, and controlled flooding increases the risk of salinity. Underground and sprinkler irrigation are more efficient in their use of water but are capital-intensive and best for high-quality plants, such as fruits and vegetables. Many nations' agricultural lands now suffer from the problem of duricrusts (crusts hardened with silica and other minerals) due to chalky soils, waterlogging, and salinity or a combination of the three: in Pakistan 73 percent of soils, in Iraq and Syria 50 percent, in Egypt 33 percent, and in Iran 15 percent.[24]

The Egyptians have irrigated from the time of the earliest dynasties, leading to urban life in the Nile valley for several millennia. The irrigation systems were based on complex canal systems and dams that the Romans adapted for their own use in Italy and Greece. At the turn of the twentieth century, Egyptian agriculture experienced a revolution based on technological change. It was primarily the result of a switch from basin irrigation, which depended on Nile flooding to soak low-lying land and provide the water needed for a single crop, to perennial irrigation using pumps, canals, and dikes. Perennial irrigation also involved greater mechanization, application of pesticides and chemical fertilizers, and use of new high-yield seed varieties. According to Paul Harrison, given the scarcity of water resources, state-sponsored research should focus on techniques and crop varieties adapted to making the most of scarce water. These crops would mature early, grow deep roots, and have the ability to withstand heat and sandblasting. Sociological research on daily use of time, farming practices, and other basic information about the area must supplement research concerning plant types.[25]

Repeated droughts and resultant famines indicate that African governments have been unable to promote technological solutions to water problems or achieve the political stability needed to deal with crisis situations. A 1983–1985 drought that hit thirty African countries required the establishment of emergency feeding camps and led to violence between ethnic groups. Some four hundred thousand nomads in Niger alone moved into the cities, putting stress on already inadequate wells and sewage systems. In Ethiopia eight million people lost their farmland and animals to drought. In Sudan the lowest Nile flow in 350 years was recorded; the water shortage left land to within thirty kilometers of the valley to dry up. When there were droughts, farmers "attempted to maintain family cereal consumption levels by expanding crops onto more marginal agricultural lands, reducing that available for livestock and overgrazing the rest."[26] All over the continent many farmers ate seed grain and sold their tools and draft animals, making it nearly impossible to commence farming again once the drought ended.

Another sign of the weakness of postcolonial states is the limited statutory infrastructure to ensure rational utilization and management of resources. According to Bondi Ogalla, the state has four primary ways to ensure access to clean water for all citizens: 1) regulation of allocation, use, and development of water resources—for example through licenses, permits, and concessions; 2) regulation of waste disposal, through procedures, prohibitions, and fines; 3) regulation of socioeconomic activities to ensure that those activities do not impinge on the state of water resources; and 4) creation of various mechanisms to assess and preempt the adverse effects development has on water resources. On the basis of analysis of laws in Zambia, Ethiopia, Ghana, Sudan, and Kenya, Ogalla determined that each government failed in some way either to promulgate effective statues or to enforce them. Discharge standards to control industrial processes or agricultural runoff are weak or non-existent. Governments rarely regulate land-use practices, and they have not defined water quality or standards. Citizens have little recourse against polluters. In rural areas the situation is always worse.[27] The weaknesses of most postcolonial states in enforcing antipollution statutes and in securing clean air and water stands in sharp distinction to the power of the state in pluralist systems.

GREEN REVOLUTION IN THE DEVELOPING WORLD: RESEARCH, BENEFITS, AND COSTS

Might green revolution agricultural technology be a solution to the problems of population pressures, difficult environmental, geological, and climatic conditions, and the endless cycle of war and poverty? Would green revolution technology made available through foreign aid have raised production? To meet a crisis in population growth and counteract stagnant agricultural production, a number of Asian countries welcomed the green revolution, thanks to which rice and wheat yields increased manyfold, although often at the cost of extensive use of chemical fertilizers, pesticides, and herbicides. In Asia one reason for the success of the revolution was that rice and wheat are staple foods. In Africa staple foods vary from region to region: in arid and semiarid regions, millet and sorghum; in the savanna, maize; and in humid and subhumid areas, cassava, yams, maize, rice, plantains, and bananas.

Couldn't research on green revolution crops appropriate to various ecosystems in Africa and to the people living in them lead to stable production? All indications are that modern agricultural technology should be adopted in many nations. The green revolution has brought huge increases in agricultural production, especially of cereals. High-yielding and hybrid varieties of vegetables and other cash crops in South Asia increased cereal, rice, and vegetable yields in the early 1960s and 1970s. Ninety percent of the wheat grown in India is now the high-yield variety. High-yield strains of rice under cultivation have spread widely throughout India, Pakistan, and Sri Lanka. Production in kilograms per hectare has more than doubled.[28]

Yet green revolution technologies failed in Africa for several reasons. One was the paucity of research on African crops. A second reason was that in many areas soils had lost most of their nutrients by the end of the twentieth century because of continuous high-yield crop production; elsewhere they had become compacted and held little water; still elsewhere, forests were cleared to make way for agriculture.[29] A third reason is that green revolution crops tend to require a lot of water. Since irrigation is needed for green revolution crops, farmers should in fact have grown traditional crops instead; indeed, they prefer traditional crops, which cost less, require less work to cultivate, and mature later than most high-yield varieties.[30] In trying out new technologies and processes, many African governments turned away from their own trained personnel, relying instead on foreigners to make decisions and looking to cash crops for export earnings, at the expense of subsistence agriculture. The promised export earnings failed to materialize. To make matters worse, few Africans were trained in environmental sciences or agricultural science or policies. The final reason, therefore, for the failure of the green revolution in Africa, B. K. Darkoh argues, was worshipping of things foreign that "served only to reinforce the existing colonial or neo-colonial mentality."[31]

In contrast to extensive agricultural research programs being developed in Europe and the United States in the late nineteenth century, little research or policy supported food crops during the colonial period in Africa, with the exception of maize in eastern and southern Africa. The International Crops Research Institute for Semi-Arid Tropics opened only in 1972 to study staple crops of the Sahel (sorghum, millet, pigeonpeas, and groundnuts). The International Institute of Tropical Agriculture (IITA) was founded in 1967, but it opened three years later in Ibadan, Nigeria, and a third of its positions were still unfilled in 1980. The IITA programs on cassava (together with those of the cassava program of the International Center for Tropical Agriculture in Cali, Colombia) are extensive and important, given the fact that cassava provides roughly two hundred million people in Africa with more than half their daily calorie requirements. The IITA has developed improved cassava varieties that are disease- and pest-resistant, low in cyanide content, drought-resistant, early-maturing, and high-yielding: they yield 50 percent more than local varieties. Hundreds of researchers and extension agents have helped introduce the varieties throughout Africa's cassava belt. Researchers from the IITA also promoted a technique to make fifty plants from each parent cassava, instead of ten stakes as before. The researchers have worked on harvesting machines that reduce processing time and labor significantly. Finally, scientists have exported cassava flour know-how to ten other countries. Madagascar, Nigeria, Tanzania, and Uganda have begun using high-quality cassava flour as the raw material for such processed secondary products as biscuits and noodles. Also being considered are new techniques to produce yams in standard shapes, to increase possibility of processing them by mechanical means.[32]

Some scientists believe that comprehensive agricultural research will not suffice. Unless farmers themselves were included through an extension service to diffuse both appropriate technology and the knowledge to use it, this would be "knowledge in a vacuum." The extension agents must see peasants as willing partners, not close-minded and backward farmers. In Kenya, in promoting cash crops somewhat successfully, developing sorely needed programs for soil conservation, and providing appropriate technical training, officials supported agricultural extension and research on a more significant scale than elsewhere in Africa: one agent for every 310 to 700 farmers. For the most part, however, African extension services pursued high-yield monocultures and expensive chemical products that came from abroad and hence required large commitments of human capital, without training and support at the levels required. In the long term, successful agriculture in developing nations may require more emphasis on local economic resources and indigenous techniques to increase soil fertility and water retention.[33]

Instead, research has supported the pursuit of chemical-intensive and water-demanding green revolution crops, including such export crops as coffee and cotton. This choice exposed farmers to the risk that their single crop would fail and to price fluctuations in markets far away. The policy tied them in to the purchase of costly seed, herbicides, and fertilizers and concentrated farmland into large tracts, as bankruptcies soared. In Sudan, Kenya, and elsewhere, wealthy businessmen and other well-connected individuals acquired land that had once belonged to peasants who had produced staples but that had become available when those farmers failed. The well-to-do forced peasants off the land to establish export plantations.[34] In Kenya the result has been the concentration of political and economic power in urban centers, at the expense of the periphery where 90 percent of the population resides, and the concomitant social displacement.[35]

In addition to the negative social impact that green revolution technology had on traditional farming, great health and environmental costs have been connected with exposure to pesticides and overuse of fertilizers in virtually all locales. Until late in the twentieth century, officials ignored these costs or had little systematic data about them. Since environmental protection laws in less-developed nations tend to be toothless or lack staff with the power to enforce them, levels of chemical use are higher than in developed nations. Further, not only the land but the health of agricultural laborers suffers. Susan Andreatta observed a terrifying example of the over-use of chemicals during fieldwork in 1994–1995 in Antigua, Bar-bados, and St. Vincent, where overspraying and drift threatened worker safety. Producers rarely offered workers such protective clothing as respirators or gloves because the items were too costly, and in any case workers would rarely have used anything so cum-bersome in the humid climate. She confirmed with extension agents the proliferation of medical problems among owners and la-borers alike from biocide poisoning. Streams and rivers have been polluted, especially during rainy season. Fish kills and pollution of the municipal water supply and outlying wells also resulted. Yet only twelve extension agents are on hand to advise thirteen thou-sand banana growers in the St. Vincent Banana Growers' Associa-tion on correct pesticide use.[36]

Whether on plantations from years past or on the small holdings that grew out of them after World War II, the agricultural econo-mies of Antigua, Barbados, and St. Vincent are geared toward monocultures of arrowroot, cocoa, coffee, indigo, spices, and sugar for export. In the search for "blemish-free" products, they have turned to "widespread use of insecticides, herbicides, and fungi-cides . . . to compete against pests and to maintain a place in the world market." The pests include rodents, insects, and weeds. The chemicals include many on the U.S. EPA list of those restricted be-cause of their harmful effects on humans and the environment and some on the EPA's canceled list. Among them are organophos-phates, paraquat, 2,4-D and others that no doubt re-enter the United States as residue on bananas, mangoes, avocados, and other products.[37]

In Asia, the average annual fertilizer use more than doubled in one decade (1975 to 1985) to ninety-three kilograms per hectare in 1985. (Worldwide use of chemical fertilizers increased almost 250 percent between 1966 and 1986.) In South Asia, fertilizer use on rice increased sevenfold in the 1960s and 1970s. In India, use of nitrogen fertilizer grew forty-one times between 1961 and 1996. Data on pesticide use are inconsistent but indicate vast increases similar to those for chemical fertilizers. Further, the use of biocides on pests has decimated natural predators of many of those pests. The result everywhere has been pollution, deterioration of the land, proliferation of pests and diseases, and a decline in human and animal health. Ultimately, the productivity of land declines as soils are gradually poisoned, and farmers use more and more fertilizers and pesticides to keep up.[38]

Green revolution nitrates have led to eutrophication (increased richness in mineral and chemical content of water and algae blooms). The damage to wildlife has been extensive, but few studies have been done to document the full extent. Birds, especially entire populations of rare cranes and storks, have been killed. Fish have disappeared from freshwater lakes and rivers. In the Chittagong and Durgapur districts of Bangladesh, fish production in paddies declined by 60 to 75 percent in the 1990s, after the green revolution.[39] Can Rachel Carson's warnings in *Silent Spring* be ignored in postcolonial regimes?

Instead, in a situation reminiscent of the 1960s in the United States at the point when Rachel Carson warned about the impact on humans (from long-term exposure or acute exposure of field-workers) of excessive use of chemical fertilizers and biocides, data about the carcinogenic or mutagenic effects on human health remain hard to come by in postcolonial states. This lack of public information once again indicates the danger to the public of having weak or poorly funded government bureaucracies to mediate the complex interplay of modern technology, social structures, and environmental change. No government organization at any level systematically gathers or analyzes data concerning hospitalizations, exposure to chemicals, or deaths. Public health officials must rely on anecdotal information in considering the impact of green revolution technology. But evidence of great human costs, including pesticide-related deaths, is slowly accumulating. In India, some 20 to 50 percent of wells contain nitrate levels higher than World Health Organization (WHO) limits. In the Kuttanad area of India, frequent cancers occur among rice cultivators. The many short-term illnesses give rise to costs in hospitalization, treatment, laboratory work, travel, and replacement labor.[40] As Australian economist Clivo Wilson concludes, "Present agricultural practices, despite producing record yields using large quantities of chemical inputs, are unsustainable and diametrically at odds with the definition of sustainable development espoused by the World Commission on Environment and Development."[41]

Those who see technology as a panacea for development problems hold out the hope that biotechnological techniques connected with genetic engineering of new plants and animals are the key to the next green revolution. But if the past is any guide, these biotechnologies may add to, not solve, the problems of social dislocation, migration, and environmental degradation because profit will be the major motive, not assistance to traditional communities, and MNCs, not local villagers, will be the greatest beneficiaries. (Keep in mind the impact on local communities of concentration of farms into agribusinesses in the United States. Is there a difference between the impact in developing nations and in the United States?) Nations that purchase biotechnology will probably also buy into increasing indebtedness and new crops that put local producers at a competitive disadvantage. The result will be the creation of scientific-technical colonies reliant on Western science and technology. It is unclear why "pathbreaking technology" that is based on gene engineering, plant cell fusion for hybridization, animal embryology, and xeno-combinations should be any different in their socioenvironmental impact from high-yield varieties. Such multinational corporations as Monsanto, Chevron, Pfizer, Sandoz, CIBA-Geigy, Celanese, Hoechst, and W. R. Grace rarely transfer know-how to developing nations, let alone communities of farmers, but rather hold patents and other proprietary information closely. And most likely, they will focus on cash crops and other commodities, not food necessities.[42] Of course, should MNCs not seek profit through their best products?

In addition to pursuing cash crops at the expense of food crops, political leaders have sought to develop cattle ranching for export markets. Just as in Brazil, where World Bank–financed ranching caused significant environmental and social problems, so in African nations ranching was no panacea for development. After the mid-1970s, in part with funds from International Fund for Agricultural Development (IFAD), an agency of the United Nations established in 1977 to follow through on the recommendations of the 1974 World Food Conference to increase food production and self-sufficiency, Botswana began to expand cattle production. Funds from IFAD were directed toward Botswana's poorest farmers, to assist them in increasing production of basic food grains (sorghum and maize) and legumes and sunflowers, in order to achieve self-sufficiency and improve income distribution. The program was successful in getting Botswana's poorest farmers to use newer implements, but data indicated no clear pattern of increased production and therefore no rationale for continuing the program. It may be that the government was more interested in seeing cattle production increase. The European Union gave preferential treatment to Botswana cattle producers to import up to twenty thousand tons of beef annually. Botswana's beef processing accounts for 80 percent of its agricultural output, and 95 percent is exported. But Fantu Cheru describes how the Botswana Cattle Project transformed the fragile Kalahari Desert and Okavango Delta into desert wasteland and killed off herds of endangered megafauna.[43] As in Brazil, most ranchers were absentee farmers who used ranches simply as places to graze livestock and generate export earnings, not to manage land for the future or to provide employment for local residents. The ranchers cut down trees in pursuit of more pastureland.

After a few years the soils on which grass the cattle graze had lost nutrients. In a word, no less than high-yield crops, ranches are capital-intensive and destructive to local ecosystems, and they destroy traditional ways of life and self-reliant forms of agriculture.[44]

HYDROELECTRICITY IN THE INDUSTRIALIZING WORLD

Might modern hydroelectric power stations with reservoirs to store water and irrigation systems to utilize it provide a way out of a cycle of drought, crisis, and poverty? Or, is capital-intensive hydroelectricity an inappropriate solution for cash-strapped countries with traditional socioeconomic structures? And what are the environmental costs of hydroelectricity? The evidence would indicate that hydroelectricity, like the green revolution, creates a large number of social, economic, and environmental problems that must be balanced against the benefits of water storage, flood control, and power generation they provide.

Many observers argue that rapid technological change in post-colonial regimes, such as that associated with hydroelectricity, leads to ecological havoc, political upheaval, economic hardship, and mass human migrations. The migrations result from relocation and displacement of farmers, peasants, and indigenous people who have lived and worked on the floodplains for generations, and put strains on resources and institutions in new regions—countries or cities where the immigrants congregate to eke out some kind of existence. In part because of Haiti's embrace of hydroelectricity, for example, hundreds of thousands of Haitians have migrated to other Caribbean countries, the United States, and Canada, by way first of urban slums in Haiti.

The case of the Peligre Dam in Haiti pointedly illustrates how weak governments that seek out large-scale technological solutions to problems of development have failed to consider adequately both the benefits and the costs of projects. They have excluded citizens at every step of the technology assessment process, and their engineers and officials have lacked the resources to produce proper environmental impact statements. The weak governments have allowed powerful businesspeople and other elites to reap many of the short-term financial benefits at the expense of the poorest members of society.

Haiti is a mountainous country with most of the land on steeply sloping terrain, only a third of it arable. The soil is particularly susceptible to erosion during hurricane season. The Peligre hydroelectric power station and irrigation facility enhanced local agricultural production in the valleys downstream. But agriculture above the dam was destroyed, in part through logging of tropical forests in the highlands, where roads were built to facilitate dam construction, factors that accelerated erosion of farmlands and forests and increased sedimentation problems. Further, the average silting in the Peligre reservoir was three times that in the design estimates, and the reservoir rapidly lost 50 percent of its holding capacity. A massive outmigration resulted, creating squatter settlements on the edge of Port-au-Prince and district capitals.[45]

Because of laws and regulations that favored them, absentee landowners gained control of much of the irrigated land that became available through the Peligre project. According to Philip Howard, they enriched themselves through rents and repression, forcing poor laborers to give up farming for work in the cities. The scarcity of land led to violent conflicts and accelerated rural poverty. Rapid modernization of agriculture in the valley increased the area devoted to rice cultivation, but without creating indigenous skills to maintain the irrigation system. Without an extension service or specialists' input, the irrigation schemes fell apart and soils became increasingly saline.[46] Once again, as with the green revolution, political elites made decisions to allocate millions of dollars to build large-scale technologies as if they could be placed anywhere successfully but neglected to provide requisite institutional support to integrate them into the fabric of society: extension services, local education and public health programs, job retraining, or resettlement costs.

In Africa, similarly, dams and large-scale irrigation projects have taken a significant environmental and social toll, in the form of resettlement, growing class bias, and new health hazards. In many coastal regions of Africa, where rivers exhibit gentle slope and low head, dams permit brackish water to penetrate estuaries, destroying fisheries. Local people have been nearly powerless to oppose hydroelectric projects, although in the case of the Kafin Zaki Dam in Nigeria they were able to stop construction, against political interests in the capital and the wishes of international aid organizations.[47]

The Kumadugu River's Yobe basin in northeastern Nigeria covers about 9 percent of the country's land, or eighty-five thousand square kilometers. The area lost 38 percent of its land to desertification in the 1980s and 1990s. Some argue that human activities—bush burning, overgrazing, and increased use of mechanical methods of land clearing—brought about this change. Kole Ahmed Shettima counters that the government, through its sponsorship of large-scale nature transformation projects, accelerated the environmental degradation of the Yobe basin. Nigeria's oil boom and rapid urbanization significantly increased demand for urban water supply and irrigation. This increase led government planners to push the creation of reservoirs in the Yobe basin. By the late 1990s twenty-two dams were either under construction or proposed in the basin, all of which were built on the promise of reducing flooding, but also to supply water for the oil industry. Officials also garnered support for the Kafin Zaki project on the basis of appeals to nationalist rhetoric and "developmentalism." They promised a secure national food supply, as opposed to reliance on imports.[48]

Canadian and United States international development agencies promoted the Kafin Zaki Dam and the accompanying eighty-four-thousand-hectare irrigation system, largely in response to a drought in the Sahel in 1972–1974, but also to prevent Soviet engineers from gaining a foothold in a country with extensive oil reserves. Shettima points out that political leaders also seized upon misleading data indicating that the impact of the dam on downstream users would be minimal. Once again, the absence of long-term data prevented optimal dam design. The leaders accepted the claims of experts that the dam would permit them to control water in an environmentally sound manner through controlled releases and to protect wetlands. Yet various NGOs, the Nigerian Conservation Foundation, the Borno and Yobe state governments, the Nigerian Federal Environmental Protection Agency, and many in the international community opposed the project. These organizations successfully argued that downstream benefits of controlled releases were overestimated and uncertain. They pointed out that past experience with dams built in Nigeria showed in fact great loss of wetlands and fauna sanctuaries. Dams had destroyed ecosystems along the Kumadugu River floodplain, on which millions of people depended for traditional agriculture. Further, the dams had high capital costs, and the irrigation systems produced low yields, had high management costs, and put small farmers at a distinct disadvantage.[49]

Nigerian academics and leaders had hoped that the Kafin Zaki Dam would serve the same role as the Kainji Dam on the Niger River above the coastal port of Jebba. The Kainji Dam, commissioned on February 15, 1969, was a major investment project to underwrite the economic development of Nigeria. Its roots date to British colonial engineering studies endorsed by local and regional government and as such was a symbol of Western approaches to economic development that may not have been appropriate for a postcolonial regime. By 1951 the Electricity Corporation of Nigeria, which was staffed by Nigerian scientists and bureaucrats who had been trained abroad, had expressed interest in the project and touted its potential benefits for power generation, navigation, agriculture, fishing, and tourism. A decade latter an engineering report approved the project. In 1962 the federal government set up the Niger Dams Authority to develop the river's potential, along with that of the Kaduna River, a major tributary. The Kainji Dam cost $262 million. Its reservoir is 137 kilometers long and nearly 1250 square kilometers and covered 360 square kilometers of cattle-grazing land. Construction required the resettlement of 44,000 people from 203 villages and led to the loss of roughly 14,500 hectares of farmland. Further, flood control was less effective than promised and in fact has led to a contraction of agriculture downstream, while fishing has been damaged, in spite of extensive expenditures to protect the fisheries.[50]

These types of large-scale, state-sponsored projects—many of which rely on Western aid, loans, and technical expertise to proceed—continue to gain supporters, even as they grow more costly, and by inundating larger and larger areas, have submerged family homes, towns, historical sites, and productive floodplain farmland. One such dam was the Aswan High Dam in Egypt.

Egyptian leaders under the nationalist president Gamal Abdel Nasser determined to build a dam as a symbol of national achievement no less imposing than the pyramids. They sought to build a high dam to store water under Egyptian control, thereby avoiding the complications of water rights negotiations that would be involved in any upstream international project (the Nile also flows through or drains portions of Sudan, Ethiopia, Zaire, and other countries). Playing the USSR and United States off against each other, Egypt secured Soviet financing and engineering. The first chief engineer on the Aswan project had worked as engineer on the Kuibyshev Dam on the Volga, at that time the world's largest, and its head designer had cut his teeth on the White Sea–Baltic (Belomor) Canal, Stalin's first major project to use slave labor.[51] The dam was completed in mid-1970, is 3.8 kilometers across the top and 111 meters high, and formed a reservoir 500 kilometers long, to support year-round agriculture, municipal purposes, and industry. Rated at 2,100 MW, it operates at roughly two-thirds capacity.[52]

The significant environmental costs of the Aswan High Dam have proven the dam's opponents right in all too many cases. The dam traps silt upstream, leading to fewer nutrients flowing downstream, and hence to diminished agricultural yields and eventually to greater reliance on chemical fertilizers. Other effects have been higher soil salinity, erosion of river bed and banks, coastal erosion due to absence of new silt deposits at the river's mouth, and destruction of fisheries by the increasingly brackish waters upstream. As for the Aswan High Dam, the reservoir's depths became a dead zone, weeds growing in along shorelines impeded water flow and navigation, and the reclaimed land was half the amount forecast.[53]

Another significant cost was inundation of Nubian history—the destruction of monuments and archaeological sites between the first and the third cataracts of the Nile River. After Vittorino Veronese, director-general of UNESCO, issued an appeal, the Egyptian government attempted to save some of the most valuable monuments, after first surveying the area and then excavating the site. The archaeologists removed twenty monuments from the Egyptian part of Nubia and four monuments from the Sudan. Others were merely identified and documented. The hydrological engineers and electricity producers were ultimately victorious, and most local historical sites were flooded forever.[54]

Yet the proponents of hydroelectric power stations can also point to the facts to argue that, in many cases, the benefits of construction far exceed the costs. One scholar argued that the Aswan High Dam represented neither "a sinister modern break with Arcadian tradition" nor a fiasco, but rather a crucial and successful effort to improve living standards for a growing population. He pointed out that 96 percent of Egypt's citizens lived within 2.5 percent of the country's total irrigated green area. This area required stable and regular irrigation, for the Nile has a sixth the Mississippi's flow (only 84 billion cubic meters annually), but 80 percent of it in three months (August, September and October). The reservoirs store water for year-round use. People had irrigated in the Nile basin for centuries, so the Aswan's irrigation systems were familiar technology. This was not a case with the international science that supplanted local knowledge. When Egypt's population doubled between 1897 and the end of World War II, and again in the next thirty years, the dam eliminated the danger of a devastating flood and secured an adequate food supply.[55]

Many of the costs of modern hydropower and irrigation projects are unanticipated, appearing only after debate has ceased. They include transmission of disease vectors, microbes, and parasites along new waterways into new regions, and along the routes of migration and resettlement. Ill-planned resettlement brings changes in diet and customs and destruction of traditional ways of life. Resettled people usually end up in crowded conditions and filth, in either unfinished settlements or slums, in which diseases, infections, flies, and mosquitoes flourish. An epidemic infection linked with irrigation canals is schistosomiasis, a snail-borne parasitic disease that afflicts some 175 million persons in the world. Critics of the Aswan High Dam predicted outbreaks of schistosomiasis; now thousands of children along the Nile suffer from liver problems connected with the disease. They defecate and urinate in running water, which is considered clean and appropriate. Before, in the dry season in Upper Egypt the land dried out and most of the snails died off. Infection rates have risen from about 5 percent to 85 percent in some places. Elsewhere Nigerians and Ghanaians have been blinded by onchocerciasis and Sudanese have visceral leishmaniasis, parasitic diseases exacerbated by dam construction. Dams have forced more than three hundred thousand people to abandon familiar farmland and resettle under difficult circumstances.[56]

Colonialism in India also led to far-reaching changes in society and environment. Subsistence agriculture gave way in many regions to manufacturing and the formation of markets for finished commodities that accelerated a breakdown in traditional communities. Formerly, community members had cooperated in production, but when markets became the focal point for organizing and controlling extensive resources, atomization of communities often resulted.

Around 1860 Britain emerged as the world leader in deforestation. Having devastated its own woods and those of Ireland, South Africa, and the northeastern United States, it turned to the East India Company, which razed Indian teak plantations, and not only to supply the British navy and merchant marine, but to demonstrate British political control. Clearing the trees created more land for cash-crop agriculture and allowed the expansion of mining industries. The building of a railroad from the 1850s on accelerated deforestation. As in Brazil with its highways in the second half of the twentieth century, many regions of India were "laid bare" when the railways penetrated into the hinterlands, both because railroads opened land to development and because they triggered an insatiable demand for wood, necessary for ties and wagons. Swaths of land were cleared and settled as thousands upon thousands of kilometers of railroads were laid.[57]

The British eventually recognized that they had triggered a crisis of deforestation, especially of the teak, sal, and deodar used for sleeping cars. The colonizers created an imperial forest department in 1864, with the help of German forestry experts to designate reserves, and established a state monopoly through the Indian Forest Act of 1865. The act pushed peasants—and their common-law claim to forests—aside. Subsequent laws fixed the proprietary right of the British state and the elimination of any community control. As demand increased, so did British control over the land, from more than 36,000 square kilometers in 1878 to almost 211,000 square kilometers of reserved forests and 8,500 of protected forests in 1890. These lumber resources were invaluable to England during

World War I, when U.S. and French supplies were unavailable and fellings shifted to remote corners of the Himalayas and the densest forests of the Western Ghats.[58]

The costs of British colonization of Indian forests were especially significant for hunter-gatherers and itinerant farmers. Individuals and Indian communities were excluded from forest management both socially and physically. Before British commercial exploitation, forest products were limited to pepper, cardamom, and ivory, commodities whose extraction did not affect the ecology of the forest or traditional practice, and to hunting. With the establishment of state control, hunting was made illegal and even access to the forest was denied. Tribes disappeared or moved on. The use of the plow and wage labor replaced both ecologically sounder slash-and-burn methods and the practice of letting some land lie fallow. Settled cultivation ensued.[59]

Indian peoples protested British imperialist policies until the establishment of independence in 1948. By restricting access to traditional sources of raw materials, the forestry department contributed to the decline of artisanal industry—for example, that using bamboo in construction, basket-weaving, furniture, and instruments. Iron production based on charcoal smelting disappeared when the advent of heavy taxes on furnaces and diminished supplies of charcoal destroyed the industry. And many other destructive British practices provoked rebellion. Gandhi led the Indian National Congress in the struggle for independence by initiating nationwide campaigns in defiance of forest regulations in the 1920s and 1930s. Direct, sustained opposition to state forest management was pronounced in a number of districts, including present-day Uttar Pradesh and Tehri Garhwal.[60]

Gandhi's approach to economic stability had been to emphasize local skills, communities, and indigenous technology. The spinning wheel on the Indian national flag symbolizes all these, as evidenced in the effort to maintain a local textile industry. Gandhi led a march to the ocean, where he engaged in the production of sea salt, which could help Indians avoid British salt taxes, and drove the point of self-sufficiency home.[61] This was a clear attempt to privilege local knowledge over Western science.

By contrast, Jawaharlal Nehru, Gandhi's successor, adopted big technology as a means of solving India's "backwardness" and poverty. Nehru was not alone in his enthusiasm for big projects. The artifacts of science, technology, and engineering have frequently fulfilled such ideological purposes as providing a sense of legitimacy to the regimes that promote them. National leaders often speak about hydroelectricity, highways, skyscrapers, and even the conquest of space as signs of their countries' superiority over the nations with which they compete for influence and economic might in the world. We call this ideological aspect of big science and technology the display value—that is, the ideological and social significance. Science and technology have also been successfully used both by colonial regimes, to uphold their rule, and by the leaders of the independence movements in Indonesia (Sukarno), India (Nehru), China (Mao), Ghana (Nkrumah) and Egypt (Nasser), to foster rapid technological advance as part of the fight for freedom and the nation-building enterprise. These leaders shared the belief that research and development focused on big technology could transform their predominantly agrarian countries into industrial ones overnight. This approach, of course, led to a kind of technological dependence (again) on industrial technologies from the advanced, rich countries, from which the emergent nations needed to acquire such modern know-how and technology as green revolution cash crops, hydroelectric power stations, and irrigation systems.[62]

The Indian government adopted this approach under Nehru. In 1958 the government sanctioned the Scientific Policy Resolution, which stated clearly that the increasing prosperity of the masses depended upon technology, raw materials, and capital, with technology being "the most important." Not surprisingly, in the 1950s Indian leaders turned to steel, coal, electric power, petroleum, chemicals, and other "classic industries" of the first industrialized nations. Eventually, India developed a significant nuclear program for peaceful and military uses, even while programs to promote poverty abatement, agriculture, and literacy languished. The country's leaders drew up five-year plans to demonstrate the rationality of its efforts. Unlike leaders of nineteenth-century nations, Nehru also advanced a rural and community development strategy. This was based on multipurpose hydroprojects, including irrigation systems, which Nehru called "temples of modern India."[63]

The "temples" created infrastructure, but left the poor disenfranchised and powerless. People were ousted from their homes and moved to other land, which was usually of lower agricultural value. They fell deeper into poverty, an indication that big technology was not a panacea for India. In the 1970s, therefore, the government turned to a green revolution strategy. Granted, yields increased greatly, but socioeconomic inequalities also increased, and food became one of India's most energy-intensive economic activities, simultaneously creating a rural market for consumer goods and exacerbating rural-urban differences. While the contribution of agriculture to net GDP decreased from 50 percent in 1951 to 37 percent in 1985, the percentage of the workforce in agriculture decreased only from 67.5 to 63.5 percent. If one takes population growth into account, the absolute number of persons employed in agriculture grew 20 percent. Investment in the urban industrial sector created only forty million new jobs. Roughly only 180 million of India's total population of 800 million really benefited from the temples of planned development.[64] We have already considered the significant increase in the amounts of chemical fertilizers and biocides used in India.

Because of uncertain benefits and the burden on the residents of the floodplains who were removed before its inundation by the waters backing up behind a dam, local residents in democratic India have protested vigorously against big projects. But usually, because of national interests and international expertise, the projects go ahead, seemingly under their own momentum. In India, Tehri in the western Himalayas became an object of desire of the Tehri Hydro Development Corporation, which had employed Soviet technical advisers. Soviet hydrologists were renowned for their unwillingness to "go slow" on any project. Their goal was to construct what was then the largest dam in Asia, at 244 meters high. When the engineers arrived, the nearest airport and railway stations were four hours away by bus. Most towns and villages along the route had no electricity. The buses passed through valleys where agriculture was on the wane for want of simple technology—like pumps. Pumps would have cost rupees; Tehri cost hundreds of millions of rupees.[65]

The Tehri project had long percolated in the minds of promoters. The Geological Survey of India had first identified the valley as a possible dam site just after independence in 1949, recommended construction in 1969, and gained approval in 1972. Progress at Tehri initially stalled because the leaders of the state of Uttar Pradesh, in which it was to be built, were unsuccessful in gaining financing for the megaproject. They could have found budgetary support for smaller mills and generators that would more cheaply and immediately have met the desperate need for electricity in the valley. Instead, the officials canceled smaller projects so that Tehri could go ahead. In reality, local needs were of little concern, for the plan was to transmit electricity to the state grid and urban users 250 kilometers away, and water for irrigation and municipal supply in Delhi.[66]

When local representatives failed in local and national courts to stop construction, the protests grew, especially through the Chipko ("tree huggers'") movement. The demonstrators conducted hunger strikes in the 1980s to halt the dam and pointed out that building a dam in a seismically active area would put at serious risk the two hundred thousand people living downstream. The Indian government hired experts who discounted dangers of earthquakes, experts who had been consultants on earlier stages of the project.[67] Unfortunately for the Chipko movement, early in 2001 the hydroelectric power company commenced construction on a major aspect of the Tehri dam, for the first time in nearly two decades. The 2,400-MW power station will open in the early twenty-first century, after thousands of square kilometers are inundated behind the dam; final testing of power generation facilities has begun.

Among thirty large dams planned for the Narmada River, the Sardar Sarovar Dam is the largest, with a proposed height of 163 meters and a reservoir 214 kilometers long. The hydropower station will produce 1,200 MW. The dam has generated great opposition because it will displace as many as 320,000 people. The Indian government claims that the dam would end drought in the state of Gujarat. Gujarat leaders are so wedded to seeing science as a panacea for poverty that they founded Science City (a series of government-supported research institutes dedicated to uniting science with regional economy), an institution of unknown benefit to farmers in the region. Opponents organized *satyagraha* (nonviolence in the Gandhian tradition) sites at Domkhedi and Jalsindhi, which were totally submerged in September 2002. Scores of people were arrested. Villages, crops, homes, and cattle were destroyed, and laws regarding resettlement were violated.[68] Of course, the officials of the Sardar Sarovar Marmada Nigam Company, the quasi-public organization behind the rebuilding of the Narmada River basin, refer to irrigation and flood control as central reasons to build dams, which will supposedly overcome the problems of irregular rainfall and the penetration of increasingly saline water into coastal regions and fight the regular droughts in the Gurjurat region.[69]

There are "hydro" success stories. The state-implemented Tyefu Irrigation System rejuvenated the Fish River, a major confluence on the eastern cape of the Republic of South Africa. The Fish River flowed only at certain times of the year, and at other times water lay in large saline pools. In the 1970s the government rebuilt the river by feeding it with water from other catchments, so that year-round corn production under sprinklers is possible. The system is only four hundred hectares but has enabled many men, mostly Rharabe Xhosa, to return from the cities to which they had migrated for jobs when agriculture failed.[70]

But all too often postcolonial regimes remained the object of large-scale development projects with significant environmental and social costs, projects that were pursued because the absence of a democratic tradition precluded substantive discussion or successful opposition, either by workers, peasants, indigenous people, or the national experts who usually bought into the projects. For example, toward the end of the twentieth century several nations of Southeast Asia also adopted the Soviet-American template for hydroelectricity. This template involved the construction of large-scale hydropower stations to produce significant benefits for lowland and urban inhabitants in the form of electricity, irrigation, and crops, but at the expense of upland ethnic minorities and ecosystems.

Thailand, Laos, Vietnam, and to some extent Burma pursued large dam projects as a way to push their economies into the international spotlight and as part of an agenda of regional integration. They focused mostly on the Mekong River basin and its many tributaries. One approach to hydrodevelopment in Southeast Asia is the so-called Mekong Cascade, similar to schema of stepped reservoirs in the Columbia and Volga River basins. Another focus is in Laos, where tributaries contribute two-fifths of the Mekong River's water and the steep topography promises very high hydroelectric potential. Dam construction, which accelerated in the 1960s but really took off after the Vietnam War ended, ultimately involved the resettlement of some 250,000 people, mostly ethnic minorities and the disruption of their lives. Existing dams have damaged forests, interrupted natural river regimes, changed flood regimes, permitted saline intrusion, reduced the flow of silt and nutrients, and lowered fish catches throughout the region. Some local opposition, the growing costs, and economic slowdowns stalled these projects until the 1990s. Then some sixty projects were advanced in Southeast Asia, with twenty-three contracts or memoranda of understanding signed, which will lead to the resettlement of many people. At least six intra- and interbasin transfer projects have also left the engineers' tables. The proposed 3,200-MW Hoa Binh Dam in Vietnam is typical of Soviet-era (and now Russian-based) megaprojects that involve a huge reservoir. In this case, to create the 230-kilometer-long reservoir, engineers will flood 400 square kilometers and displace about sixty thousand people, most of them ethnic minorities.[71]

Have large-scale projects been appropriate for the developing world? What are the potential environmental and social costs associated with them? What goals should the nations pursue? If the goals are to eradicate poverty and ensure basic living standards, to slow the rates of rural emigration by stabilizing the quality of life in the countryside, to focus on the cultural as well as the socioeconomic well-being of the people, and to abandon the urban mindset, then modern technological approaches have had mixed results on many counts. If the goals are to secure stable economic growth, to create organic linkages between town and country, to avoid social disruption, and to limit environmental change, then support for traditional, agriculturally based artisanal technologies that draw on renewable sources of energy (solar, biomass, or hybrid) and for educational programs seem more appropriate.[72]

WASTE DISPOSAL: FROM FIRST- TO THIRD-WORLD COUNTRIES?

Such big science and engineering projects as green revolution agriculture and hydroelectric power stations had an especially dark side. Many postcolonial regimes went deeply into debt to fund the projects and felt compelled to import hazardous waste and noxious industry to pay down the debt and secure employment for their citizens. By the late 1980s some thirty to forty-five million tons of toxic waste were traded every year. More than half of it went to non-OECD countries, and a fifth of it went to the third world. By the end of the twentieth century the export of toxic waste was a thriving industry. Countries that imported waste generated significant earnings from it. At least nine African nations, including Benin, Gabon, Nigeria, Sierra Leone, South Africa, and Zimbabwe have accepted hazardous waste for storage. This solution enabled Western companies to avoid the costs of compliance with legally mandated standards for disposal of these various materials at home.

The highly toxic compounds stored include cyanide, asbestos, lead, and persistent organic compounds. The African governments lacked the expertise and state-of-the-art facilities to control the materials properly. Often waste was simply dumped in leaking containers—or no containers at all. In Koko, Nigeria, PCBs and asbestos have leaked from drums. Incinerator ash has been dumped in the open air in Guinea and Haiti. In many cases, the wastes, because they are mixed with organic materials, ended up in fields used for agriculture and thus entered the food chain. The employees who handled the hazardous materials lacked appropriate equipment, clothing, and training and paid the price in exposure to deadly poisons.[73]

An alliance of NGOs and several nations called for the banning of trade in hazardous waste. As a result, the Basel convention was adopted by representatives of dozens of nations in March 1989. The Basel Convention on the Control of Transboundary Movements of Hazardous Wastes and Their Disposal (to give it its full name) resulted in the regulation of trade in waste, not in an outright ban. The convention urged states to handle their own waste close to home, not to export it. The convention also limited waste trade to countries that were parties to the convention and prohibited the export of waste to Antarctica and the export of radioactive waste generally. Yet the convention was vague on how to stop illegal shipments and on what constituted hazardous waste and what was "environmentally sound." No discussion of liability was included. Further, bilateral waste deals were permitted, and the largest "waste makers"—the United States, Canada, and Germany—refused to sign the treaty. The United Nations Environment Program called for a full ban on toxic waste. Sixty countries agreed, but six opposed the ban (the United States, Australia, Japan, Canada, the United Kingdom, and Germany). Self-interest almost always carries the day; the United States generates 85 percent of the world's total hazardous waste, the European Community another 5 to 7 percent. Effectively, then, the convention is not in force.[74]

The Basil convention required that exporters notify recipients of shipments and receive approval beforehand. Some 2.6 million metric tons of hazardous waste may have been shipped to the south or east between 1989 and 1994. Between 1990 and 1993 alone, Australia, Canada, Germany, the United Kingdom, and the United States shipped another 5.4 million metric tons of hazardous waste, but under the permitted category—and loophole—of "recycled" material, to thirteen Asian countries. John Ovink doubts that the importation and processing of "recycled waste" will benefit the receiving countries. One reason is that the receiving countries usually lack monitoring procedures, let alone facilities to recycle the waste in an environmentally sound fashion.[75]

Recognizing how they had increasingly become dumping grounds for the industrial countries, the African nations signed the Bamako convention in 1991, to prohibit the importation of hazardous waste from industrialized nations. But economic pressures have forced the African nations' hand.[76] Not only wastes are moving south. The United Nations estimates that between 20 and 50 percent of all foreign direct investment in developing countries is in "pollution-intensive industries"—chemical, pulp and paper, metals, coal, and petroleum. According to Joshua Karliner, factories of transnational corporations in southern-tier nations contribute directly and indirectly to about 50 percent of all emissions of greenhouse gases.[77]

Several incidents involving the export of dangerous industrial production and waste from Europe and America to the developing world have achieved horrible notoriety. On December 2, 1984, a storage tank at a pesticide plant in Bhopal, India, exploded, releasing a cloud of methyl isocyanate gas. The gas, spreading over the homes and hovels of nearby residents, choked and poisoned them. Six thousand died within a week. According to some estimates, sixteen thousand people have died in the long run. The city of Philadelphia loaded toxic ash from its municipal incinerators onto a ship in August 1986. Officials intended to dump it at a site in the Caribbean. After a year and a half of seeking some country that would accept the waste legally, the ship probably dumped the waste in the Indian Ocean.

Among industrialized nations, disposal of their dangerous radioactive waste in developing nations has been openly discussed. For years the United States, the Soviet Union, France, Britain, and China dumped radioactive waste at sea in steel and concrete caskets. The Soviets surreptitiously continued this practice into the 1980s. They stored it on land in steel tanks that began to leak within a few years, although scientists assured us that such waste would not leak for thousands of years. Because of the intractability of the problem, the costs involved, and the desire of nuclear powers to put their wastes in someone else's yard, extensive discussion has been devoted to burying waste in third-world countries. For example, British scientists explored geological deposits in Sudan.[78] The notion of transporting radioactive waste by ship, truck, or train from one country to another is exceedingly disturbing, especially in consideration of the fact that terrorists may target those shipments.

TECHNOLOGY AS A TOOL OF FOREIGN RELATIONS

Thus far, we have considered the impact of the colonial experience on the interrelationship of science, the environment, and the state. Because of the weakness of many postcolonial states, and in some cases their corrupt nature, the governments of these nations have had a difficult time dealing with the environmental legacy of colonialism and in adopting innovative technologies in agriculture, industry, and other sectors of the economy that are environmentally sound. In many cases, postcolonial states face shortages of water and other natural resources that make escape from the cycle of poverty and environmental degradation a great challenge. When leaders looked to the promise of high-yield crops, the results were often disappointing because the foundation in research was lacking, as were extension services to inform peasants how to care for the crops properly. Postcolonial regimes also lacked the legal framework to regulate dangerous production processes or protect natural resources. All too often, decisions to pursue industrialization, electrification, and urbanization ignored the true cost to the poor, the disadvantaged, and the minority nationalities within their borders. In many cases the nations became dumping grounds for the West's waste.

In a word, technology became of major tool of a foreign policy in the relationship between northern- and southern-tier nations, and postcolonial nations have had great difficulty in using this tool to their advantage. Aid and technological-assistance projects in Africa have failed at much higher rate than in other regions of the world. Institution building—creating scientific, educational, and social service organizations to employ modern technology properly—have moved at a snail's pace. Many major projects (for agriculture, livestock, hydroelectricity, roads, refineries, and mines) actually had negative economic returns. How much of this situation was the legacy of the colonial period, when Europeans exploited Africa as a source of raw materials? Why does Africa remain an exporter of raw materials, while industrialization and modernization of agriculture lag behind? Does the neocolonial legacy of the green revolution make Africa dependent on exports of minerals and cash crops? Are huge hydropower stations merely white elephants?

Some analysts argue that technology itself is "a major instrument of domination in international relations, especially in North-South relations." According to this view, modern technologies, when introduced in developing countries, rather than solving various economic problems, in fact increased technological dependence. These countries usually lacked the ability to generate, adapt, and use technologies, for they did not have a research-and-development apparatus or people trained to undertake these tasks. Instead, technology distorted their development patterns. Technologies have high capital costs and usually benefit urban residents, while leading to significant social, economic, and environmental upheaval. Technology enters developing nations through investment in various projects, particularly in the extractive industries and other raw-materials sectors, or in the form of licensing, patent deals, or management contracts. Technology may take the form of entire turnkey plants, know-how, or aid programs. Hence, technology is not merely a thing in itself but is rather embodied in the economic and political will of its providers.[79]

The weakness of most African and other postcolonial states enabled MNCs to gain undue influence over many national economic decisions that have an irreversible impact on the environment. Joshua Karliner argues that MNCs gained great power in determining the direction of the world economy through the World Trade Organization, the European Union, and other international organizations. Ninety percent of MNCs are in OECD countries. The General Motors and Ford Motor Corporations together "exceed the combined GDP of all sub-Saharan Africa, with 51 of the largest 100 economies in the world being corporations." According to Karliner, this economic power gave multinational corporations the ability to determine "where factories will be built, which forests will be cut, minerals extracted, crops harvested and rivers dammed." The corporations' wealth enabled them to buy up privatized ore, fish, forest, oil, and other resources—and other corporations. They used subcontractors rather than affiliates to undertake some of the most hazardous activities, to avoid direct responsibility (a case in point was Bhopal). Multinational high technology, such as large-scale factory fishing fleets, has wiped out local fishing communities, and lucrative vegetable and fruit export farming has replaced land once used for subsistence farming. Such firms as Chevron, Shell, and Mobil developed close relationships with authoritarian regimes (for instance, Nigeria).[80] Workers in industries established abroad by MNCs run the risk of exposure to dangerous conditions and hazardous materials. Wages remain low, and workers have no organizing rights.

In underdeveloped countries, most of them former colonies, indigenous science and engineering have lagged behind. As noted earlier, research and extension services for agriculture tend to be weak in these nations. The total world share of research and development among less-developed countries (LDCs) is 3 percent; of that, much is concentrated in India, Argentina, Brazil, and Mexico. Ninety percent of the world's scientists and engineers work in developed countries. Often, where there appears to be "knowledge-based" industry, the reality is different. Ruben Berrios, a Peruvian social scientist, characterizes some nations in which workers assemble computer circuits as "microelectronic colonies" of the industrial world, not truly developed nations.[81] Some observers therefore argue that aid to industrializing countries ought to focus more on promoting science and education to develop their own scientific institutions and not so exclusively on enabling them to purchase modern technology as such.

While highly critical of the place of high technology in North-South relations, Berrios acknowledges that LDCs must share the blame for how technology is employed in industrializing nations. These nations lack a coherent technology policy, let alone a scientific research establishment capable of providing the appropriate channels for the introduction of new technologies. In other cases, economic activity has revolved around "the indiscriminate purchase of consumption technologies," such as automobiles, by the elites, not around job-creating technologies.[82]

In addition to MNCs, direct foreign investment, and foreign aid, industrializing nations receive funds for development through such organizations as the World Bank. The World Bank was conceived during World War II, and its initial focus was rebuilding Europe after the war. Its first loan, of $250 million, was to France in 1947 for postwar reconstruction. But while reconstruction remains a focus to help nations deal with natural disasters, war, and transition, the World Bank has for many decades also focused on reducing poverty, a noble and desirable goal. Yet criticism of its programs grew during the 1970s and 1980s because the kinds of large-scale projects that the bank tended to approve to fight poverty underestimated or ignored environmental and social costs. The projects, in addition to being highly capital-intensive, disrupted the environment and destabilized indigenous social structures. Moreover, in some cases bank managers ignored the coercive measures recipients employed in their pursuit of development. Some infamous cases involved funding for cattle ranches and extractive industries in Brazil and hydropower stations throughout the world.

In the late 1970s the World Bank began to pay more attention to environmental concerns in evaluating projects. In 1981, the World Bank recognized, at least on paper, that ecological concerns must be considered in approving projects, along with measures of economic growth, and it adopted its first environmental policy (twelve other guidelines have been added since then). But even with the creation of an environmental department within the bank, representatives of many NGOs, of the media, and of the scholarly community dismissed the bank's attempts to "green itself, for too many of its projects were failing." Only in the 1990s did it begin to require an environmental-impact assessment for all projects.

Environmental considerations have sociocultural aspects as well, one of which is gender. For example, on the basis of her analysis of the Indian Sardar Sarovar hydroelectricity project, Priya Kurian argues that World Bank environmental assessments, in focusing on issues of rationality and efficiency, paid inadequate attention to human cultural diversity. Significant environmental degradation and human rights abuses accompanied the bank-funded Polonoroeste project in Brazil, transmigration in Indonesia, and the Sardar Sarovar dam in India. In the last case, women were excluded from local decisions, two hundred thousand people were displaced, and only men, not women, were compensated for their resettlement. Project managers have begun to consider nutrition, basic education, sanitation, water supply, and housing in evaluating projects, along with population planning, energy conservation, health care, pesticide use, pollution control, and conservation, but these efforts must be expanded.[83]

WHITHER TECHNOLOGY IN THE DEVELOPING WORLD?

Agricultural, medical, energy, and other technologies have played an important role in southern-tier nations modernizing their economies, and in enabling their citizens to gain access to public health, transport, and power production technologies. But as I point out in this chapter, the acceptance of technology often requires adaptation of existing political, social, and other structures to ensure that the technology works properly. As the cases of high-yield agricultural varieties and hydropower indicate, the pursuit of modern technology is fraught with difficulties for traditional societies, and perhaps particularly for those in which the state is weak. Whether this is so because of the legacy of colonialism or because of other factors remains an issue for discussion.

There must be greater discussion of the many paths to employ technology in aid projects and the ways to estimate their environmental and social costs accurately. This discussion should consider not only the various kinds of aid, but also the level of aid that wealthy nations offer, since only Denmark, Norway, the Netherlands, Luxemburg, and Sweden devote at least .7% of their GDP to foreign aid, as agreed to at the 1992 Rio conference on environment and development. Some planners and engineers seek to anticipate all technical, climatic, and social problems connected with their projects. They have tried to estimate the costs of disruption to traditional ways of life and the benefits that citizens derive from them. Yet balancing costs and benefits in advance may be impossible, since most engineers and policy makers consider the successful completion of a given project the most important indicator of benefit.

For their part, when they offer aid, the wealthy nations rarely provide it in forms that are easily assimilated. Given that the receiving nations have only small scientific and engineering establishments, it makes little sense to provide primarily technological aid without the simultaneous transfer of know-how. Other forms of "technology" that could be immediately valuable—for example, licenses to produce antidiarrheal drugs or AIDS drugs at low cost—rarely find their way into aid packages. In the case of antidiarrheal drugs, they do not generate big earnings for pharmaceutical companies, and in the case of AIDS drugs, the companies wish to recoup drug development costs by charging higher prices and not permitting early production of generic versions of their drugs.

To what extent have postcolonial administrations embraced

Western technology and attitudes? In most cases government experts, advisers, bureaucrats, and planners wish to end pastoralism and other agricultural practices that they see as outdated and inefficient. In most cases they consider big "state farms and large-scale commercial holdings" as an answer to food problems. More and more wealth and power has become centered in the cities, where civil servants, trade unions, military officers, business interests, and politicians are located. These people often see peasants only as a source of taxes or as obstacles to progress.[84]

In turning to environmental concerns, national governments that control wildlife, range lands, and natural forests often ignore or prevent local community control. There are few wardens, foresters, or range managers to protect lands, and they have small budgets for enforcement. As Paul Harrison notes, each family, at the mercy of environmental conditions and political desiderata, exploits resources as quickly as possible in the "tragedy of the commons" par excellence.[85] Weak states may simply be incapable of defending the environmental interests of their citizens. The greatest hope for the development of an appropriate relationship between the state, technology, and the environment lies in increased aid to less developed nations, in the form not of weapons but of appropriate technology and educational programs, and in promulgation of treaties based on the recognition that a clean environment is a human right, as I discuss in Chapter 4.

CHAPTER FOUR

BIODIVERSITY, SUSTAINABILITY, AND TECHNOLOGY IN THE TWENTY-FIRST CENTURY

We want the maximum good per person; but what is good? To one person it is wilderness, to another it is ski lodges for thousands. To one it is estuaries to nourish ducks for hunters to shoot; to another it is factory land. Comparing one good with another is, we usually say, impossible because goods are incommensurable. Incommensurables cannot be compared.

—Garrett Hardin

To this point, we have discussed the ways in which polity, economic system, and worldview affect the establishment of environmental policies, concepts about ecology, and the development and diffusion of technologies that have an environmental impact. I have suggested that pluralist regimes generally have been more successful at weighing the costs and benefits of unregulated economic growth, mitigating environmental degradation, and balancing the interests of citizens than have authoritarian, colonial, and postcolonial regimes. Still, all of them, under the influence of the Enlightenment worldview or other ideologies of progress, have been attracted to large-scale approaches to harness, cultivate, harvest, and process natural resources: hydroelectric power stations, aquacultural systems, industrial forests, agribusinesses, and extractive industries. These have tended to be more destructive to natural ecosystems and traditional socioeconomic relationships than small-scale approaches. Large-scale technologies usually require determined political action to come into existence. For example, building a huge hydroelectric station and irrigation complex will make it necessary to appropriate extensive stretches of land and to resettle tens of thousands of people, often those least able to protect their own interests, not to mention submerge thousands of kilometers of land and the homes, churches, cemeteries, and historical sites on that land. We have also discussed the fact that in pluralist regimes, where public participation in technology assessment is an accepted aspect of the policy process, damage to the environment and society from rapacious development or harm from the inadequacy of statutes to protect the environment occurs less frequently than in authoritarian regimes or in developing nations.

Another important issue we have considered is the way in which worldview shaped the transformation of nature, power production, resource extraction, waste disposal, and silvi-, aqua-, and agriculture in the twentieth century. At the beginning of the twenty-first century most scientists and engineers, most policy makers, indeed most of the world's citizens, believe strongly in the power of modern science and technology to help manage natural resources. In countries that have fully embraced the Western scientific and engineering ethos concerning the desirability and feasibility of control over or even improvement on nature, the Enlightenment worldview will probably continue to bear fruit. The standard of living in those countries is the highest in human history, infant mortality has decreased, life expectancy has increased, diets have improved, and health care is better.

We have also considered the ways in which the effort to transform nature and in many respects to put it on an industrial footing, is based on a change in worldview. This change involved the certainty among leaders, scientists, and ordinary citizens that humans not only could but ought to strive to control and improve on nature. Humans were more powerful than nature itself, so the thinking goes. Were problems to be encountered, a technological solution to those problems could be found—even if the problem was technological in origin. A crucial factor in this change was the central role of the modern nation-state in fostering environmental change. It did so through support for study of natural resources, expeditions, systematic gathering of data, establishment of national laboratories, and creation of bureaucracies and agencies to promote resource development and geoengineering. States undertook resource study and development toward the ends of military strength, economic stability, and improvements in public health. Governments without exception passed laws to assist businesses and industries toward these ends. The result, in many places, has been the achievement of very high standards of living, including consumption of more foods, fuels, luxuries, and services. Many people, however, have no understanding of where the goods and services originate that contribute to the urban lifestyle.

It is also clear that the levels of consumption to which many people are accustomed cannot persist indefinitely. Forests have fallen rapidly to developers, logging firms, and paper companies. Trawlers have destroyed fishing grounds. Problems associated with handling and storage of hazardous materials continue to grow. Greenhouse gases have already had such a significant impact on climate that only a few citizens of the globe continue to ignore the evidence of global warming in favor of a philosophy of "economic growth at all costs." Other problems of pollution and hazardous waste disposal remain without solution.

In this chapter I discuss the changing range of environmental issues that confront governments and their scientists in the post–cold war world and outline the major environmental and health issues at the beginning of the twenty-first century.

INTERNATIONAL CONVENTIONS AND TREATIES

International bilateral and multilateral treaties are one approach to regulating behaviors and technologies with the aim of limiting and reducing global environmental degradation. Given that nations will not be a party to treaties which require changes in national regulations, or behaviors they are unwilling to embrace within their own borders, success in using them has been mixed. Considering experience with several treaties and conventions directed at environmental problems will make it easier to understand the limitations of treaty negotiation and enforcement. We have already discussed the Basel convention on hazardous waste. We shall now consider carbon dioxide (CO_2) and other greenhouse gases and global warming and chlorofluorocarbons (CFCs).

Water, methane, CO_2, ozone, and other gases (including oxides of nitrogen and CFCs) in the atmosphere absorb heat. Carbon dioxide is the gas primarily responsible for global warming. Its concentration in the atmosphere has rapidly increased since the industrial revolution. It is produced primarily by burning of fossil fuels. (Because trees absorb CO_2, deforestation has increased concentrations of the gas in the atmosphere.) Given the importance of coal, gas, and oil for heating, electrical energy production, and internal-combustion engines, it would be impossible to wean industrial societies away from the use of fossil fuels. More efficient combustion processes, alternative energy sources, and conservation have led to modest, almost fleeting improvements in the situation. The United States—nearly alone among nations—steadfastly insists that conservation is all but impossible, and that more fossil fuel reserves must be located and tapped to ensure future economic growth. In fact, automobiles are the major culprit in the production of CO_2 in the United States, and they have become the primary source of pollution in virtually all nations of the world. The world's economies produce around 5,500 million tons of CO_2 annually.

Greenhouse gases have a number of surprising sources. The

growing demand for beef throughout the world also contributes to production of the gases, in this case methane. Such South American and African nations as Brazil, which see the technology of cattle ranches as a way to stimulate modernization of their agricultural sectors and to generate export earnings and jobs, have felled vast forests and offered extensive subsidies, loans, and tax incentives to ranchers to meet worldwide demand. Cattle release large quantities of methane. Methane is also produced by rice fields, landfills, and fossil fuel use. Methane is twenty more times effective in trapping heat in the atmosphere than CO_2. There are 1.3 billion cows in the world today. Most of them belong to big agribusinesses and ranches. A cow releases 600 liters of methane daily, so one cow produces 219,000 liters annually and the world's cows release 300,000 billion liters annually. Factory farms, which have grown significantly in size and number as the world's appetite for beef grows, clearly have significant environmental costs.

The history of global warming indicates how challenging it has been to comprehend the ways in which industrialization, technology, and environmental impact are interrelated. Scientists have observed—and debated the causes and extent of—global warming for well over a century. Some specialists saw the modest yet troubling temperature increases as reflecting the cyclical nature of average global temperatures and cited evidence drawn from analyses of ice cores to bolster that position. For others, even taking the cycles into consideration, the rise in temperature of 1.5 degrees or more over fifty years has left no doubt that human activity, and in particular the rapid increase in fossil fuel use from the 1870s onward, has filled the atmosphere with greenhouse gases such as CO_2 that trap solar heat. The views on global warming are a good example of a dispute among experts and of the paradox that seemingly objective data produced by truth-seeking scientists often provoke widely differing interpretations about the causes of the phenomenon at hand. In this case, most of the scientific community now agrees that hu-

man activities are the chief cause of global warming, that calls for
further study to verify global warming are efforts to postpone difficult but necessary regulatory action, and that international as well
as national responses are required. Finally, the history of global
warming reveals the crucial role of the state during the twentieth
century in mediating, regulating, or promoting technological development, as policy makers attempt to weigh the benefits of economic activities (factory production that provides jobs, goods, and
services) and its costs (long-term environmental degradation, emphysema, and so on).

As James Fleming has written, one century after the discovery of
the stratosphere, only five decades after the invention of CFCs, and
only two decades after chemists warned about destructive nature of
chlorine and other compounds, we understand that human activities have damaged the ozone layer in the atmosphere. After
centurylong study, including now computerized modeling of the
atmosphere, we have recognized the possibly catastrophic impact of
industrial pollution. As far back as the eighteenth century, David
Hume explicitly linked European agricultural habits and clearing of
forests with the rise in temperatures. By the beginning of the nineteenth century, a partnership of government and science had begun
to monitor the climate systematically. For example, the Smith-
sonian Institution, the Army Medical Department, and the U.S.
Coastal Survey were all involved in gathering and analyzing data. In
1896 Svante Arrhenius published a paper on increasing levels of
CO_2 in the atmosphere. Although his work was concerned with
explaining the ice ages, not with alerting us to some impending
greenhouse effect, Fleming tells us, people wrongly consider
Arrhenius the first person to have formulated idea of global warming. The fear of the role of anthropogenic (human-produced) carbon dioxide in climate change is largely a post–World War II phenomenon, based both on scientific research and on growing public
concern that the world was getting warmer. The International Geo-

physical Year (1957) yielded several long-term studies critical to this growing awareness. During the 1970s, cold war fears that the Soviet Union was working on weather control, and growing concern within intelligence agencies that climate change, whether inadvertent or malicious, might hurt the United States, exacerbated worries about global warming and stimulated further study.[1] Satellites and computers have enabled extensive gathering and analysis of global climate data that for most scholars leave no doubt about the impact of anthropogenic carbon dioxide. By the 1990s, the governments of the world recognized the need to address the problem.

According to the accord on greenhouse gases reached in Kyoto, Japan, in 1997, the world's nations agreed to cut energy use. In the United States the use of coal, oil, and other fossil fuels was to be cut by more than 33 percent and fossil fuel emissions were to be cut to 7 percent below 1990 levels by the 2012 deadline. The European Union was to cut fossil fuel emissions to 8 percent below 1990 levels, and Japan to 6 percent below. Supporters of the treaty argued that global warming would cause catastrophic flooding, deadly storms, and tropical plagues, and the evidence linking these events and global warming is irrefutable. Opponents insist that the Kyoto accord was based on faulty science and worry that the mandatory cutbacks in energy use will cripple the U.S. economy, lead to a slowdown in production, and leave millions of workers jobless as factories shut down and move to developing nations like China that impose few, if any, regulations on pollution. Some people believe that fossil fuel emissions reductions at this level will result in a lower standards of living for consumers and a long-term reduction in economic growth. Yet even after rejecting the accord, the United States has lost millions of jobs, and now that the United States has rejected the accord, what hope is there for slowing global warming?

The Kyoto accord is a superb example of how difficult it is for governments to confront global environmental problems. The accord was drafted by more than 160 nations and applies to 34 industrial nations. China and India refused to accept any emissions cuts, arguing, not without reason, that cuts would prevent them from achieving the living standards to which Western nations are accustomed. The conference became deadlocked when a bloc of developing nations, known as the Group of 77 (G-77), sought to block a U.S.-backed proposal that would have permitted poorer countries to join the accord voluntarily and accept mandatory targets and timetables for their own emissions. The G-77 acquiesced to a U.S. proposal that would allow American companies to receive credit toward pollution-cutting targets by making deals to help clean up foreign factories and other pollution sources. United Nations negotiators dropped a demand that the top 25 developing nations come on board with their own series of targets and timetables by 2005, even though China, Brazil, and India alone will account for half the world's greenhouse gases by early in the next century. China flatly refused to join, announcing that it would accept no limits within the next fifty years.

The United States also gave up both its request to exempt U.S. military training and overseas operations from emissions reductions and its opposition to the so-called E.U. bubble (a plan that would allow massive increases in greenhouse gases in Portugal, Spain, Greece, and Ireland to be offset by reductions in Germany and Britain). Americans also offered to make steeper cuts in U.S. emissions—to reduce emissions to 7 percent below 1990 levels instead of cutting pollution only to 1990 levels, according to its initial plan. Because of opposition to what it sees as a poorly negotiated treaty that will hurt the U.S. economy, the United States under President George W. Bush rejected the Kyoto accord out of hand and called for additional study to understand the sources of global warming. In the meantime the administration attempted to weaken the Clean Air Act and to slow only the *increase* in greenhouse gas emissions. Ultimately, the history of global warming has yet to be written. Are any treaties of global significance success stories?

A major success story is the Rio convention of 1992. In the effort to address such problems as global warming, to support sustainable development, to reduce resource use, and to preserve biodiversity, representatives of 172 nations met in Rio de Janeiro, Brazil, in June 1992, at the United Nations Conference on Environment and Development (UNCED), often known as the Rio convention. More than a hundred heads of state or government attended, an indication of worldwide recognition that a global approach was required to address intractable environmental problems. As a preliminary to the meeting, a U.N. commission on environment and development, led by Gro Harlem Brundtland of Norway, had put forward the concept of sustainable development as an alternative to an approach based simply on economic growth. The commission defined sustainable growth as growth that "meets the needs of the present without compromising the ability of future generations to meet their own needs." The 1987 Brundtland report provided further impetus to call the Rio summit, to establish guidelines for development that would support socioeconomic and environmental stability and would lay the foundation for cooperation on the issue between the developing and the more industrialized countries.

The participants agreed to the so-called Agenda 21, the Rio Declaration on Environment and Development, a statement of forest principles, the United Nations Framework Convention on Climate Change, and the United Nations Convention on Biological Diversity, the last two of which required future ratification. Many of the points in the agreements were simply common sense—for example, calling for new patterns of production to limit such toxic components as lead in gasoline and hazardous waste, seeking alternative sources of energy to replace fossil fuels that cause global warming, and promoting effective new public transportation systems, in order to reduce vehicular emissions, congestion in cities, and health problems caused by air pollution.

Agenda 21 was a highly political document. It contained detailed proposals for action in social and economic areas (such as combating poverty, changing patterns of production and consumption, and addressing demographic dynamics), and for conserving and managing the natural resources that are the basis for life—protecting the atmosphere, oceans, and biodiversity, preventing deforestation, and promoting sustainable agriculture. Signatories were urged to involve important groups that had been ignored in previous discussions of sustainable development—especially women and children, but also trade unions, farmers, and indigenous peoples, local authorities, and representatives of NGOs. Agenda 21 also recognized environmental rights as a human right. Agenda 21 states that human beings, who are at the center of concerns for sustainable development, are entitled to a healthy and productive life in harmony with nature. It further states that scientific uncertainty should not delay adoption of measures to prevent environmental degradation, but many states, precisely, use "scientific uncertainty" in some cases as a pretext to delay action. The document maintains the sovereign right of states to exploit their own resources, but not to cause damage to the environment of other states. It calls for redistribution of wealth, a provision that wealthy nations may reject. In many ways, therefore, Agenda 21 implicitly and explicitly recognizes that while the introduction of modern technology can be a grand human achievement, it often carries negative, unintended consequences for traditional social structures, women, children, and ecosystems.

In addition to Agenda 21, the governments adopted the Rio Declaration on Environment and Development, a series of principles defining the rights and responsibilities of states, and the statement of forest principles to promote sustainable management of forests worldwide. The participants agreed to open for future ratification two new conventions: the United Nations Framework Convention on Climate Change and the United Nations Convention on Biological Diversity. The nonbinding statement of forest principles was the first global consensus reached on forests. It urged developed countries in particular to make an effort to "green the world" through reforestation and forest conservation. It permitted states to develop forests according to their socioeconomic needs, preferably in keeping with national sustainable development policies, but recognized that little could be done to prevent any one nation from embarking on or maintaining patterns of profligate resource use and development that had global ramifications (for instance, rain forest exploitation in Brazil and South Asia, burning coal of low calorific value in China, or automobile use in the United States).

To make matters more confusing, a number of nations' leaders subsequently retreated from key principles of Agenda 21. For example, whereas President George H. W. Bush agreed to the agenda, his son George W. Bush rejected the so-called precautionary principle of "common but differentiated responsibilities," which states that the nations most responsible for an environmental problem—and those with the greatest capacity to act on it—should take the lead in addressing it. Instead, at the beginning of the twenty-first century, the United States developed an energy policy that calls for a substantial increase in the use of fossil fuels and other nonrenewable resources and was willing to address the contribution of automobiles to global warming, clearly the major source of greenhouse gases, only through voluntary measures, not through research and development of alternatives or through regulation. To put it bluntly, so wedded was the Bush administration to the technology of internal-combustion engines and fossil fuel boilers that it refused to seek any policy to reduce greenhouse gas emissions, or to sign the Kyoto protocol.[2]

At the World Summit on Sustainable Development ("Rio plus Ten") in Johannesburg, South Africa, in summer 2002, representatives of the Bush administration revealed instead a "new approach" to reduce poverty and protect the environment through partnerships with businesses, international groups, and friendly countries. Examples of the initiatives range from protecting the Congo rain forest to providing clean water and energy to the poor, plans that U.S. officials believed demonstrated U.S. commitment to sustainable development. Another partnership involved growing bird-friendly "shade coffee" in a partnership with Starbucks. Some of the small-scale partnerships have been successful—for example, one that provides 2,500 subsistence farmers in Kenya with honey-producing beehives. Others are larger, such as a joint effort by the United Nations, any given national government, the Natural Resources Defense Council, and British Petroleum to phase out leaded and high-sulfur gasoline and diesel fuel.

According to some observers, the main problem with the new approach was that it was no approach at all. The United States has historically devoted a smaller percentage of its budget to foreign aid in support of such "new approaches"—or any aid, for that matter—than has any other industrialized country. The United States spends roughly .1 percent of its GDP on foreign aid. Granted, foreign aid is not necessarily a good indicator of a nation's policy on sustainable development. Rather, the key indicators are the level of resource extraction, recognition of which resources are renewable, and production of waste. Yet many people recognize that aid to poor countries can help ease both poverty and environmental degradation.[3] Unfortunately, the level of spending is roughly the same for U.S. trade in weapons as for U.S. foreign aid for these kinds of constructive programs. Further, officials wish to base many of the partnership programs on crops that are chemical- and water-intensive and important in the export market (and therefore not in the least "local" or sustainable) and that enable U.S. firms to compete in all markets. This approach has had a particularly insidious impact on the agricultural economies of developing nations. Paradoxically, farm subsidies in the United States—a multibillion dollar nonmarket solution to a nonexistent problem of underproduction—have given U.S. firms a significant (and government-subsidized) advantage in world food markets that has hurt the agricultural sectors in developing nations. These direct and indirect subsidies have led the U.S. government to consider the need for nonmarket, including high-tech, solutions and have increased foreign aid payments to those countries. Yet levels of foreign aid commensurate with the problem have not been forthcoming.

Much greater success has been achieved in negotiating an agreement to protect the ozone layer that surrounds the earth and protects the environment from ultraviolet rays that can lead to cancer and genetic mutations. In 1982 scientists first detected massive holes in the ozone layer over Antarctica, which were growing rapidly. The scientists recognized that CFCs were a major contributor to this hole. Use of CFCs had become widespread in aerosol sprays, refrigerators and air conditioners, foam for insulation, solvents, fire extinguishers, and many industrial processes. After carbon dioxide, methane, and nitrous oxides, CFCs are one of the most important gases contributing to global warming—even though their concentrations are measured in the parts per trillion, as opposed to parts per billion for carbon dioxide and methane—and CFCs are entirely of human origin. And CFCs destroy ozone (itself a greenhouse gas). The clear dangers of CFCs and alarm over the vivid image of a "hole" in the earth's protective layer led government representatives to meet in 1986 in Montreal, where they quickly agreed to ban CFCs from use by 1996 (the Montreal protocol on CFCs). The availability of alternatives to CFCs, in contrast to the absence of alternatives to fossil fuels, may have been the single most important factor in securing rapid agreement in Montreal. Since the signing of the protocol there has been evidence of improvement, although the ozone hole continues to grow. In what might, again, be viewed as an inappropriate action, in 2003 the administration of George W. Bush asked for, but failed to receive, an exemption for the production and use of methyl bromide, a pesticide, which contributes to ozone depletion.

In addition to the Montreal protocol, several other international treaties have successfully promoted resource management practices commensurate with the availability and renewability of resources and with demand for them. Why have some agreements been successful, while others have failed? It may be that where the economic costs of noncompliance with a treaty, protocol, or agreement are immediate, local, and significant, success is more likely. One such treaty is the Law of the Sea, passed at the Third United Nations Conference on the Law of the Sea (1982), which establishes that a country bordering on an ocean can claim a zone extending two hundred nautical miles beyond its shores.

The Law of the Sea promotes peaceful scientific research into sustainable fishing practices and encourages the transfer of technology to developing nations that would permit sustainable practices. The law calls for "conservation of the living resources," which involves determination of the allowable catch of the living resources in each nation's exclusive economic zone based on the best scientific evidence available, as well as proper conservation and management measures, to ensure against overexploitation. Signatories assume the obligation to maintain or restore populations of fish that are under threat within their exclusive economic zones and to cooperate with other states to ensure the proper conservation and management of living resources on the high seas. The Law of the Sea also carries several stipulations prohibiting pollution and dumping of hazardous waste. The law calls on foreign nuclear-powered ships and ships carrying nuclear or other inherently dangerous or noxious substances to respect, and to take all precautions for the safety of, the rights of nations through whose waters they pass.[4]

The Law of the Sea entered into force on July 28, 1994, and more than 130 states are party to the convention. Even in the face of the rapacious trawling of ground and pelagic fish in areas well beyond the two-hundred-mile limit (and well beyond natural replacement levels of stocks), the two-hundred-mile limit specified by the Law of the Sea is recognized as a principle of customary (as opposed to statutory) international law. This recognition has triggered the establishment of quota, limits, that may lead to recovery of stocks of cod, tuna, and other fish and has encouraged governments to close areas to fishing within their two-hundred-mile limits.

There are, of course, limits to how much nonbinding international treaties can accomplish when short-term self-interest in profit motivates economic activity. With the establishment of the two-hundred-nautical-mile zone, such nations as Thailand whose large fishing fleets had searched far and wide for larger catches suddenly faced new problems. Thailand was the leading fishing nation in Southeast Asia in the 1970s. In 1986 it had a fleet of sixteen thousand vessels, of which 85 percent plied the Gulf of Thailand. The two-hundred-mile zone suddenly put a third of its catch in non-Thai waters; the waters near its shores could in no way produce enough fish to fill all these vessels or to meet the national need for fish for food and export. The zone created conflicts between Thailand and Vietnam, Cambodia, and Malaysia, since Thai vessels continued to fish far from Thai shores. These nations used fines, impoundment of vessels, and emprisonment of sailors to protect their waters from Thailand's fleet.[5]

This discussion of several treaties indicates that many environmental problems are global in nature and require international action. Many aspects of science, technology, and industry—among them acid rain, nuclear waste disposal, and overfishing—have transnational implications for the environment and for resource use. How successful have international environmental agreements, conventions, and treaties been in dealing with these challenges? Perhaps the most important change in the twentieth century was the recognition of transnational impact—that is, of the global environmental consequences of many human activities. Beyond that, the challenges for the state are immense. In an edited volume that examines international agreements on flora and fauna, the Montreal protocol, trade in hazardous chemicals and pesticides, air pollution, whaling, nuclear dumping, and several other issues, Kal Raustiala and David G. Victor write, "The outcomes of implementation [of multilateral treaties] are often uncertain, especially when implementation requires influencing the behavior of a larger number of social actors." Many countries take different approaches to the same problem, focusing, for example, on different sources of NO_x emissions. Some countries interpret regulations as applying to transboundary problems only, not to control of emissions or other pollution problems within the nation. Also, resolution of a growing number of international environmental problems would require significant economic and social change. The upshot is that implementation is a "demanding task" and "the means and outcomes of implementation varied and uncertain." Yet Raustiala and Victor conclude that institutions—national and international, governmental and nongovernmental—may be the best place to focus efforts to change behaviors and implement policies, since institutions are the concrete entities that are most amenable to introducing and enforcing policy.[6] There are reasons for hope, however. They include the development of the notions of biodiversity and sustainability, their transformation into a virtual call for action, and the presence of technologies that offer efficient alternatives to disruptive large-scale technological systems.

SUSTAINABILITY AND BIODIVERSITY

Nearly every reader has heard the terms *sustainability* and *biodiversity*. But what do they in fact mean? There is no simple answer, because the meaning of the terms has evolved over twenty-five years. In some cases these have become empty catchwords for complex issues. That sustainability and biodiversity are crucial concepts cannot be questioned, to judge by the way they have penetrated academic and policy-oriented literature. Major funding agencies and philanthropic foundations regularly seek to support projects that explore these ideas. The United Nations has worked since the mid-1980s to define them. The notion of international environmental security has developed concurrently with them.

I would like to propose definitions for sustainability and biodiversity in the interest of stimulating fruitful discussion. The easier of the two concepts to define is biodiversity. The concept of biodiversity has deep historical roots, stretching back to well before the time when Charles Darwin observed the rich tapestry of flora and fauna and referred in *On the Origin of Species* (1859) to "the entangled bank" of nature. By the turn of the twentieth century, botanists, entomologists, and zoologists, among others, understood the need to study the complex interrelationship of species in various ecosystems. The scientific literature of the mid-twentieth century is filled with references to biodiversity. And of course, Rachel Carson implicitly and explicitly brought to public attention concerns about the food chain and the destruction of biodiversity through overuse of chemical biocides. Yet it may have been the publication in 1988 of *Biodiversity,* edited by E. O. Wilson, that led to the elevation of biodiversity to the status of a biological concept. In *Biodiversity,* dozens of men (and only two women) offered various perspectives on biodiversity, including several on the role that technology and science might play in maintaining biodiversity, but the consensus was that "technology is not a panacea for the disease of extinction."[7]

Of course, *biodiversity* refers to the wide number of species of flora and fauna, the 751,000 known insects, 24,000 flatworms and roundworms, and 50,000 mollusks, the vertebrate and invertebrate animals, the vines, trees, grasses, and so on, that total 1.4 million different organisms.[8] But the idea of biodiversity has come to include the teleological view that all living things have value, that no species is "better" than any other, and that humans must strive to preserve ecosystems, to enable as many of them to survive as possible. Biodiversity carries the value judgment that the loss of any one species is a loss for all species. The goal of preserving biodiversity, as the word implies, is to struggle mightily to prevent the extinction of any species. The moral and aesthetic reasons for seeking to maintain biodiversity reflect the beliefs that all life is worth living and that the beauty of nature consists in its diversity. The biological concern is that the loss of any species will weaken a given ecosystem. Some people argue that humans have no right to destroy other beings, and that destructive capability is neither godlike nor desirable. Though some debate how many species go extinct annually, well-informed scientists have no doubt that the rate and number are both increasing.

Loss of biodiversity occurs because of rapid development and the resulting loss of habitat; pollution; poaching (especially of megafauna); and an intentional or unintentional introduction into an ecosystem of new species that upsets the balance because they have no natural predators.[9] Nations around the world, in the effort to preserve biodiversity, have adopted different approaches. They have limited or outlawed development in especially sensitive areas—for example, through the establishment of national parks, inviolable preserves, or protection of wetlands. As noted earlier, commons and enclosures date to the seventeenth century, and public parks and wilderness preserves to the late nineteenth century. By the end of the twentieth century, many governments had passed laws to require developers and industrialists to file environmental impact statements indicating that economic activities would not threaten fragile ecosystems. Of course, developers, fuel and ore exploration companies, ranchers, and others have sought to weaken or void laws passed to protect biodiversity, by claiming that the laws go too far, interfere with their personal property rights, or assign excessive value to nonhuman living things. In Brazil, agribusinesses fret about land set aside for reserves for indigenous people, although the amount of land the agribusinesses own and have stripped of vegetation to allow ranching on very favorable financial terms far exceeds that devoted to reserves.

The concept of sustainability has a short history, one that did not appear in scientific or social-scientific literature until the early 1980s. The notion of sustainability implies that certain ways to organize economic activity, including resource management, will preserve biodiversity. These ways are efficient, they are not resource-intensive, and they are likely to be small in scale. They may require reductions in consumption, but their supporters argue that adherence to tenets of sustainability will in no way lower the quality of life. Rather, like the Progressive-Era "gospel of efficiency" (that is, conservation), sustainability will enable present and future generations to have adequate access to natural resources and to benefit from their use. Without being framed as such, the concept of sustainability has been reflected in a number of important laws and treatises since before the 1980s. The National Environmental Protection Act of the United States (1969) recognizes the need "to foster and promote the general welfare, to create and maintain conditions under which man and nature can exist in productive harmony and fulfill the social, economic and other requirements of present and future generations." The Club of Rome study *Limits to Growth* (1972) also called for sustainable development requiring a change in patterns of consumption.

Many nations have laws to protect endangered species. A region with an obvious need to preserve biodiversity is the Amazon of Brazil, with its extensive forest, in all some six million square kilometers, an area equal in size to the combined territory of twenty-five European countries that contains roughly 30 percent of the remaining tropical rain forest in the world. The Amazon has an enormous beneficial influence on both the regional and the global climate, in that the area holds carbon stocks of around 120 billion tons. Brazil is in all likelihood the country in the world that is richest in biodiversity, with thousands of endemic species, including 131 species of mammals, nearly 200 birds, 172 reptiles, nearly 300 amphibians, and 16,500 to 18,500 plants. Many species are already threatened, including 103 birds, 15 reptiles, 5 amphibians, 12 fish, 34 invertebrates, and nearly 1,400 plants.[10] The Amazon is home to perhaps half of all the world's insect species and more than a fifth of its plant species.

Because of the emphasis placed on extractive industries, lumbering, monocultural farming, and ranching, Brazilian planners have ignored the need to conserve this natural wealth. The government has belatedly sought to develop ecotourism, promote biodiversity prospecting, and support other sustainable activities. Only in 1997 did Brazil establish its Program of Molecular Ecology for Sustainable Use of Biodiversity (called Probem), with the goal of making the Amazon region an important source of value-added products and advanced scientific know-how, especially through the use of biotechnology. Scientists now recognize the critical responsibility to develop the Amazon's wealth in a sustainable fashion, based on preservation of its endemic species, many of which have medicinal, cosmetic, and industrial value. Longtime inhabitants of the forest, and especially indigenous peoples, will also be able to benefit if they receive remuneration for their contributions to the development of new products, such as their traditional knowledge of the medicinal properties of native plant or animal species. Still, the pressure on those resources from the urban inhabitants of the coast remains strong, and every year more of the rain forest disappears.[11]

Some scholars argue that biodiversity may serve as a resource to promote economic development, while others are more measured in their evaluation of its potential either to generate income benefiting local communities or to preserve species. Julie Feinsilver asks, "Will developing countries gain competitive advantage by marketing biological resources over which they have sovereignty to pharmaceutical, agricultural and industrial firms?" Can a nation with unique flora use them to enter markets for oils (for cosmetics), herbal preparations, agricultural chemicals, and industrial enzymes? Feinsilver points out that pharmaceuticals from plants are a growing market. More than 60 percent of the world's people depend directly on plants for their medicines (through homeopathy, herbal remedies, and the like). In developed countries many of the top-selling drugs are derived from or inspired by natural sources. Taxol is perhaps the best-known example of a pharmaceutical produced from flora. Taxol comes from the Himalyan yew and has been used quite successfully in the treatment of ovarian cancer. The U.S. National Cancer Institute's Natural Products Branch has examined more than one hundred thousand plant extracts such as those from the yew since 1960.[12]

Yet to take advantage of what Feinsilver calls biodiversity prospecting, a country must have a scientific infrastructure, intellectual property laws to protect its scientists' finds, conservation areas to preserve flora and fauna, and the will among the political leaders to take advantage of this resource. Since the 1992 Rio summit, several countries with bioprospecting opportunities have passed regulations on the collection and export of biological resources, to protect them from external exploitation. Feinsilver offers the example of the Costa Rican National Institute of Biodiversity (INBio), which signed a $1-million agreement with Merck Corporation, one of the world's largest pharmaceutical companies, for biodiversity prospecting. INBio, founded in 1989, identifies flora for potential development. As part of the agreement, Merck helps train local scientists.[13] Whether biodiversity prospecting must be supported by multinational corporations and whether it truly benefits the host country remain important questions.

In the United States significant progress has been made in the effort to preserve ecosystems and endangered species within them for the enjoyment and betterment of current and future generations. Until recently, both Democratic and Republican presidents supported measures to preserve systems and species through the Antiquities Act of 1906 and the Endangered Species Act (ESA). When he signed the ESA on December 28, 1973, President Richard Nixon said, "Nothing is more priceless and more worthy of preservation than the rich array of animal life with which our country has been blessed." According to a U.S. government Web site on the ESA, the law "provides for the conservation of species which are in danger of endangerment or extinction throughout all or a significant portion of their range and the conservation of the ecosystems on which they depend." The ESA derives its power from the fact that any individual or organization may petition to have a species considered for listing under the act as endangered or threatened. Once again, we should recall the fact that pluralist regimes are more likely than others to recognize and institutionalize the civic responsibility of citizens to support environmental activities and recognize that they are as valuable as economic activities. The listing of a species through the ESA qualifies it for increased protective measures. As of 1996, 952 species were listed, with 139 proposed for listing, 179 likely to need listing, and nearly 4,000 more of deep concern to specialists.[14]

The U.S. Fish and Wildlife Service and the National Marine Fisheries Service are responsible for the listings and coordination of activities to ensure protection. An extensive scientific study based on public information and data relevant to the size and life history of the species will then be carried out. A species may be considered endangered according to five criteria: "1) present or threatened destruction, modification, or curtailment of its habitat or range; 2) over-utilization for commercial, recreational, scientific, or educational purposes; 3) disease or predation; 4) inadequacy of existing regulatory mechanisms; and 5) other natural or manmade factors affecting its continued existence." Within one year the responsible agency must make a decision on listing a species as endangered. If it determines that a species is threatened, the agency then requests additional public comments, and a final decision is made within a year.[15]

Crucial to the success of the ESA, and of concern to opponents who view the act as unnecessary or perhaps even wrongheaded, the law prohibits any consideration of the economic impact when decisions about species listings are made. This stipulation has led to confusion about listings, because many other environmental laws require or are traditionally based on an effort to quantify the costs and benefits of certain activities (cost-benefit analysis). Many aspects of the compilation of environmental impact statements involve cost-benefit analysis. For example, a project for a shopping mall may require draining wetlands, altering the path of a stream, or clearing a small woodland. The mall most likely will generate jobs for scores of people. How do we balance the costs of "losing" nature against the long-term benefits of jobs? Can we fairly amortize the costs over a long time? Many things are not quantifiable—for example, beauty, justice, and equity. And it is often difficult to compare what is more important when it comes to land use—for example, a park versus a hospital. Both sitting in a park and having access to hospital beds during time of illness are important. With the ESA, at least as initially passed, the only consideration was whether a given human activity would endanger a species or threaten it with extinction.

A little-known and rarely convened Endangered Species Committee has the discretion, however, to decide which species should be saved from extinction when an equally important project that promises economic growth would be derailed by an ESA action. The committee (often called the God Squad) consists of six high-ranking federal officials and one state resident appointed by the president. The committee can act only when asked to by a federal agency or a governor and only after it has been determined that a species is in certain peril of dying out. The God Squad has allowed projects to go on in cases where the economic costs of protecting a species have been deemed too pernicious.

An early case in which the courts determined to permit development rather than halt it in light of the ESA or the National Environmental Protection Act concerned the Tennessee-Tombigbee Waterway. The courts allowed the Army Corps of Engineers to complete the canal project, even though such species as the snail darter would be endangered by the waterworks. (The project's promoters—the corps, local businessmen and businesswomen, and powerful congresspersons—exaggerated the expected benefits of the canal for inland shipping, agriculture, and black laborers. The waterway has failed on all counts.)[16]

One international treaty on biodiversity is the Convention on International Trade in Endangered Species of Wild Flora and Fauna (CITES). The convention was drafted as a result of a resolution adopted in 1963 at a meeting of the World Conservation Union. In March 1973, representatives of eighty countries met in Washington, D.C., to approve CITES, and it entered into force in July 1975.[17] Today, it provides some degree of protection to more than thirty thousand species of animals and plants, whether they are traded as live specimens, fur coats, or dried herbs. The convention is intended prevent smuggling and trade in animal parts. These include skins and pelts for coats, hats, shoes, and handbags; rhino horns, tiger paws and bear bladders for aphrodisiacs; and ivory for carvings. In autumn 2002, CITES signatories met to approve new rules permitting transport of blood, hair, or feathers (but not reproductive tissues) from the field to laboratories back home for use in diagnostic and other genetic tests. The rules spelled out precise definitions for samples, quantities, and uses. Mexico, Brazil, and China, among others, opposed the new rules out of fearing that they would allow "uncontrolled access" to extensive genetic resources.[18] Representatives of Botswana, Namibia, and Zimbabwe also asked CITES members to permit annual quotas for ivory trade. South Africa eventually joined them. The argument was that trade in ivory stockpiled from elephants that had died naturally or ivory confiscated from poachers would bring in earnings to expand preservation areas. The representatives also claimed that this plan would satisfy world demand and put poachers out of business. But other nations strongly opposed the action, fearing that poachers would take the measure as a carte blanche to kill elephants again. The petition was dropped.

A central issue is how to balance economic growth with regulations on economic activities, resource management, and waste disposal practices to protect the environment. People differ over how to achieve steady economic growth to benefit a nation's, and the world's, citizens. Opinions differ within nations and from country to country. Disagreements also arise over what kind of government is most successful at producing sustainable economic growth. Is it one that avoids regulatory pressures on industry—for example, the requirement to cease production of greenhouse gases and other emissions, and insistence on use of the best available (often the most expensive) technology? Is it one that seeks new ways to manage resources and new technological approaches to develop them? Clearly, sustainability requires, to the extent practicable, greater reliance on such renewable resources as biomass for energy production, rather than on nonrenewable resources, such as fossil fuels. Environmentalists generally seem to agree that sustainability requires substantial reduction in consumption among the wealthy countries and better stewardship of nonrenewable resources, both in developed countries and especially in the developing world.[19] Many people also argue that the more governance and production occur on the local level, the more likely it is that sustainable patterns of life will develop. Is there some kind of technological solution to the conundrums of protecting the environment in which science, government, and civic culture can participate? I would argue that the concept of appropriate technology has provided a positive answer throughout history.

APPROPRIATE TECHNOLOGY

Some people go so far as to argue that a special kind of technology, appropriate technology (AT), inherently ensures sustainable practices. It would seem that appropriate technology cannot be capital-intensive. For example, an investment of about $400,000 in capital is required per worker in the modern automobile industry. Appropriate technologies tend to be small in scale, energy-efficient, and locally manufactured, or at least locally run and maintained. In a now classic book hailed by counterculture and mainstream economists alike, *Small Is Beautiful* (1973), Ernst Schumacher called for "technology with a human face," based on small-scale approaches that employed workers fruitfully, abandoned mass production of consumer goods where it was inappropriate, and reflected a new view of "development" that was not resource- or capital-intensive.[20] Denton Morrison developed categories for judging "hard" and "soft" technology—"soft" accurately defining what I mean by appropriate technology—which I have adapted as follows:[21]

Hard Technology	*Soft Technology*
Is capital-intensive	Is low-cost, low-capital
Depends on organized economic monocultures	Depends on small entrepreneurship, diversity
The elite benefit through large-scale organizations	Benefits more individuals; has low-income differentials
Decision making is authoritarian and centralized	Decision making is democratic and decentralized
Trickle-down innovation leaves the poor untouched	Relies on local innovation and diffusion
Brings technological unemployment, alienation, large wage differentials	Offers meaningful work, little division of labor, full employment

Motivations are profit and greed	Motivations are local needs and quality of life
Leads to profligate use of resources; human and environmental costs are greater than productivity benefits	Leads to sparing use of nonrenewable resources
Emphasizes crowded, centralized, urban lifestyle	Emphasizes decentralized lifestyle with symbiotic rural-urban relationship

Source: Adapted from Denton Morrison.

We must distinguish appropriate technology from indigenous technology, which may also be small in scale and low in capital resources. Indigenous technology—for example, the traditional agricultural and fishing systems in Africa that we have considered—is often labor-intensive, is subject to frequent failures, and may carry great environmental costs, through irrational resource use during times of crisis. Appropriate technology combines Enlightenment visions of progress with rigorous scientific understanding, reasonable capital costs, and limited impact on existing social structures.

Appropriate technology may be high technology. One such example is solar cells for production of electricity. Because of its high cost relative to fossil fuel generators (two to five times more costly per kilowatt/hour), solar power has not been widely used. Yet, according to practitioners of sustainable growth, cost should only be one factor in determination of a path of action. They believe that solar power's "renewability" is a more important consideration than its slightly higher cost. As for low-tech solutions for sustainability, they include biomass systems (for example, dung generators), drip irrigation systems, and introduction of a variety of cropping patterns for food, instead of exclusive concentration on green revolution cash crops that require significant and expensive inputs of chemicals and water. Generally speaking, appropriate technology should be based on local needs, should not be capital-intensive, and should be renewable.

What is appropriate technology for poor countries such as Bhutan, which imports most of its technology, accepts no direct investment from foreign companies, and suffers from labor shortages much of the year? What technology is appropriate, in light of Bhutan's mountainous geography and lack of roads and other infrastructure? Bhutan's citizens are ill equipped to answer these questions, given their low literacy rates (around 10 percent) and the small number of experts in the country (in all perhaps 150 trained professionals). Its one university has 150 students, of whom only a fifth focus on the sciences. The government supports only agricultural research. Bhutan's economy is small, with 4 percent of its GDP in manufacturing and mining and 13 percent in construction as of 1985. Ninety percent of the population works in agriculture.[22]

In Bhutan, therefore, appropriate technology would mesh well with the unskilled labor force and the paucity of engineers. According to Laughlin Munro, AT would also employ the smallest outlay of labor and capital, be the least harmful to the fragile mountain ecology, and contribute to economic growth and self-reliance. As noted, Vietnam, Laos, and other Southeast Asian nations have opted for huge hydropower stations enabling them to sell surplus electrical energy to generate capital. But the hydroelectricity, cement, and cash crop route is precisely the wrong one for them and for Bhutan because it flies in the face of the criteria for appropriate technology. In view of their steep, mountainous terrain, labor shortages, and small farms, Munro urges the Bhutanese to turn to "hand operated diesel- or petrol-powered tilling devices one-two person similar to those developed in Japanese hills" and mechanical threshers. Still, these devices create the problem of fuel supply for peasants, who are often far from roads (only two thousand kilometers in total length). Further, mechanization is not a panacea. It will not improve land quality in and of itself. Increasing deforestation and the expansion of agriculture into marginal lands may require Bhutan to import technologies to increase the productivity of agriculture while slowing the pace of deforestation.[23] What is appropriate technology in this case?

Are there reasonable alternatives to energy-intensive industrial agriculture? This is an important question, both because energy use for agriculture in industrialized countries is very high and because many developing countries have embraced it. Granted, high-yield varieties have produced bountiful harvests, but after repeated use these varieties require more fertilizers, insecticides, and herbicides to achieve the same yields. In addition, transport costs have risen, and development costs for hybrid seeds have grown. Green revolution rice varieties, for instance, require fifteen times more energy than alternatives. Further, the increases in output lag behind energy increments. Barnhard Glaeser and Kevin D. Phillips-Howard write that in the United States, energy use in agriculture grew threefold from 1940 to 1970, while food consumption per capita increased only slightly. They point out that energy-efficient alternatives to the green revolution exist in the manipulation of nutrient cycles, maintenance of crop diversity, biological methods of controlling pests, weeds, and erosion, use of manure rather than chemical fertilizers and of mechanical over chemical pest control, substitution of disease-resistant for high-yielding varieties, and crop rotation. Mixed cropping restricts the spread of pests and diseases, controlled burns kill ants and termites, hand-picking of insects requires no pesticides, and herbs, bark, fermented liquids, ash, and saltwater substitute adequately for pesticides and fungicides. The authors' data indicate that these alternatives can produce sufficient quantities of foods to meet demand. Glaeser and Phillips-Howard offer as an example one community in rural southeast Nigeria, without draft animals (because of the tse-tse flies) or tractors, where the inhabitants farm competently using hoes, take advantage of annual flooding, and plant small gardens near their homes which they fertilize with domestic waste.[24]

Another example of a low-cost, decentralized AT based on renewable resources is biogas generators. A major problem discussed in Chapter 3 was deforestation and pollution associated with some peasants' moving farther into the forests in search of wood for fuel often of low calorific value. Population pressures make biomass (wood, charcoal, and crop residues) more costly and scarce. Biomass is the fourth-largest source of energy in the world, contributing about 15 percent for household use—cooking, spacing heating, crop drying, brick and pottery making—and in some countries even for the manufacture of steel and rubber. In developing countries, biomass accounts for 50 percent of the energy consumed; Ethiopia derives 93 percent and Kenya 75 percent of all energy needs from biomass. The case of Kenya indicates what might be done with biogas generators.

Eighty percent of Kenya's twenty-five million people live in rural areas. Few of them have electricity, and most are poor. Kerosene and charcoal are expensive. The responsibility for going long distances to gather wood falls primarily women and children. The women's health suffers from their inhalation of smoke and soot while cooking with such low-grade fuels as animal dung and crop wastes. The results are impoverished soil, denuded forests, desertification, and high health costs.[25]

Could biogas generators running on cattle manure and urine serve to alleviate these problems? A family with four cows has access to more than eighty liters of dung and urine daily. Working with the German Technical Aid Department *(Deutsche Gesellschaft für Technische Zusammenarbeit)* of the Ministry of Energy, Kenyan agricultural specialists developed two biogas plant designs to take advantage of biomass. Some Kenyan biogas plants date back to the 1950s. They consist of vessels ranging from eight to ten cubic meters, in which the mixture of biomass and water ferments, producing methane for various cooking, lighting, power, and irrigation applications. The cost, at 1,230 to 5,400 pounds sterling, was too high, so by 1984 only about 160 units had been installed, and of these, fewer than 25 percent remained in operation. The German and Kenyan specialists discovered that well-trained artisans—a kind of extension service—were required to help farmers ensure that their units functioned properly, while the government needed to provide access to capital and materials.[26] Many people who worry about heavy reliance on the government insist that the private sector is best equipped to promote technological innovation. But as the case of biogas in Kenya shows, the poor require financial and technical assistance to take advantage of AT, from which the environment and the entire nation will benefit.

Through its Rural Energy Technology Assessment and Innovation Network (RETAIN) the Canadian government in the 1970s supported the diffusion of biogas generators in China, an important program, given China's growing reliance on highly polluting low-grade coal for energy production. The fuel source for these generators was crop residues and human and porcine waste, which were converted into methane and fertilizer. China remains one of the few nations that has successfully transferred AT "from the urban and scientific centers to the rural periphery." The Chinese government promoted the installation of huge numbers of biogas units in a short time. Yet it supported the construction with propaganda, but not with sufficient resources, and as a consequence many of the seven million biogas plants built in between 1973 and 1978 did not work. Like the government of Kenya, Chinese officials belatedly recognized the need to establish an extensive service for the units: seven hundred biogas service stations and seven thousand construction teams.[27]

It seems that AT programs often fall prey to short-term economic considerations, instead of benefiting from long-term environmental and social ones. A fall in the price of oil in real and relative terms after 1973 removed energy technologies from the top of the foreign aid policy agenda. The United States Agency for International Development (USAID) cut back on energy projects in the late 1970s and the 1980s because those projects largely failed to meet their goal of providing efficient and simple energy technologies with low capital costs. Still, volatile world oil prices have left developing nations in particular in an energy bind, and fossil fuels are nonrenewable. As Barnett points out, rural energy development is essentially a public sector activity. There is not enough profit at stake to draw firms to the countryside, and scientists and engineers tend to focus on other lines of research, so the state must be involved.[28]

There is also AT for such societies as the United States. Can an alternative be found to the American proclivity for the massive yet mass-produced homes (derisively called "McMansions") with such faux signs of opulence as front-porch columns? Since Americans tend to occupy, and therefore heat and illuminate, the largest domiciles in the world, many times larger than in any other nation on average, the simplest approach would be to build smaller houses. But Americans seem inclined to ignore moral or other suasions to build smaller homes. Perhaps only a significant increase in energy prices would encourage the construction of ecologically sound homes, although a luxury tax on any square footage above a certain amount, similar to the "gas-guzzler" tax on luxury vehicles, might have an impact. In some communities, strict building codes require Americans to build ecologically sound homes. Certain federal, state, and local codes encourage such practices. Yet there is no agreement among the codes, and since the expiration of the tax credits from the Carter era for making energy improvements to homes, no federal incentive exists to do so. But such codes—based on climate—would include requirements on the amount of insulation in walls and ceilings, on energy-efficient windows that open, on designs limiting the number of windows on the north side of the house, on the planting and maintenance of shade trees, on energy-efficient heating and air-conditioning systems (including roof fans), and on waste systems to recycle gray water for gardening. Appropriate technology for automobiles, arguably the single most profligate user of nonrenewable resources, has long been available. Manufacturers produce engines that run cleaner and use less gasoline. Plastic and composite materials lower the weight of vehicles. Engineers have developed hybrid electric-gasoline vehicles. But without prodding by the federal government through CAFE standards that require mile-per-gallon averages for a fleet to improve—something that is reasonable and feasible within the next model year—little can be done. Fortunately, state governments have statutory authority to regulate polluters like automobiles. California adopted "technology-forcing" legislation to require manufacturers to achieve higher than federal standards. The manufacturers groused, but were able to meet this standard, an indication that all that is missing is political willpower. Of course, neither home nor automobile AT can alter consumption or throwaway patterns, for to do that would require a change in worldview.

A CHANGE IN WORLDVIEW AS THE SOLUTION

This text commenced with a discussion of the importance of worldview for considering the interrelationship of technology, the environment, and the state in the twentieth century. The combination of Enlightenment thinking about the desirability of improving on nature; the rise in statecraft that incorporated this worldview as part of its ideology to provide for defense, economic growth, and public health; and the expansion of the industrial and scientific bases of society all resulted in significant improvements in access to and management of natural resources. Improvements in the quality of daily life and public health followed as well. Initial efforts at flood control, dredging of harbors, development of hybrid corn, and the like, led many scientists and policy makers to view technology as a universal solution to problems in securing and distributing resources through more or less decentralized means. In democratic societies, perhaps because of the participation of the public in the policy process, the benefits of modern technology were more evenly distributed among citizens, and the costs—including environmental degradation—were both more evenly distributed and less extensive than in authoritarian regimes. In the developing world, to a greater or lesser degree, dependence on Western engineering traditions, complex obligations to international financial institutions, and the choice by elites to pursue a path that benefited minorities led to selection of development strategies that were damaging both to traditional socioeconomic institutions and to the environment.

Many analysts have failed to consider the place of science and engineering both as root causes and as solutions to the world's ongoing environmental crisis. Unquestionably, scientists and engineers are dedicated to finding solutions to the crisis in a variety of ways. These ways include promoting energy-efficient technologies; developing more productive, less water- and energy-intensive food crops; and focusing efforts on renewable resources. Of course, scientists and engineers cannot solve what most observers agree to be the real problem: overconsumption of goods and resources in many modern economies. Citizens, especially those in North America, seem reluctant to change their consumption patterns. They consider access to resources a right, not a privilege, and certainly not the luck of the draw. Their political leaders all too often lack the courage to challenge the view that profligate use of resources is some kind of manifest destiny. Are they wedded to the belief that the "market" or "technology" will find solutions to environmental problems of their own accord, like some disembodied supernatural force? Don't citizens themselves have to make choices about the appropriate path of economic growth?

To judge by polling information, citizens in North America,

Europe, and Japan nearly without exception believe that environmental problems require immediate attention. They probably welcome open discussion about the options, risks, and costs of uninformed—if not underhanded—decision making. Perhaps political leaders are merely too beholden to short-term economic interests rather than long-term environmental ones. And all too often leaders find it easier to endorse large-scale projects for resource development, at the same time begrudging adequate support for social welfare or educational programs. To make the situation worse, once big projects commence, they are nearly impossible to stop. As Heyneman notes, project planners speak glibly of the need for interdisciplinary information to ensure against unanticipated costs. But once dams, canals, highways, or reactors have been approved, planners view them largely as engineering problems, whose construction is based on economic considerations, and whose effects on public health and social or cultural values they can ignore or regard as less important.[29]

While many governments have made significant progress in promoting sustainable development, preserving biodiversity, and addressing century-old problems of waste, too many others have taken halting steps at best. The world faces an environmental crisis at the beginning of the third millennium that grows worse day by day. Throughout the world the regions inhabited by indigenous peoples have disappeared, as ever-growing hungry populations exploit shrinking forest and water resources, and urban populations demand immediate satisfaction of their needs from resources that lie far away. The habitat of many species has rapidly dwindled. Numerous species are already extinct. Disputes over resources may lead nations to the brink of war. For example, in the Middle East, Israel uses three to five times per capita more water than Palestine or Jordan, and water is the crucial factor for life in the Middle East.

In a classic article published in 1968, Garrett Hardin warned the citizens of the world about the "tragedy of the commons." A commons is any resource that we use in the belief that it belongs to everyone. It is shared land or water that we take both because we assume it to be unlimited and because we fear that if we do not use it, others will. We use it to maximize our benefits. Yet we usually destroy the commons through unregulated use. Hardin argues that we treat many of the earth's resources as if they were part of the commons—to bad effect. He also notes that pollution is connected with the tragedy of the commons. He writes, "Here it is not a question of taking something out of the commons, but of putting something in—sewage, or chemical, radioactive, and heat wastes into water; noxious and dangerous fumes into the air; and distracting and unpleasant advertising signs into the line of sight. The calculations of utility are much the same as before. The rational man finds that his share of the cost of the wastes he discharges into the commons is less than the cost of purifying his wastes before releasing them. Since this is true for everyone, we are locked into a system of 'fouling our own nest,' so long as we behave only as independent, rational, free enterprises."[30]

Hardin's solutions to the tragedy of the commons disturbed a number of people. He argued: "A finite world can support only a finite population; therefore, population growth must eventually equal zero." He therefore called for population-control measures. Since population growth is highest in less developed countries, it would fall to the inhabitants of those countries to limit their populations. This opened Hardin to the charge of eugenics or racism, since the brunt of population-control measures would fall on the nonwhite people of the world. Leach and Mearns argue that Hardin's argument is neo-Malthusian and has ignored a true commons situation in which local institutions facilitate cooperation among users, so that resources are managed sustainably. The authors argue that the tragedies are Hardin's invention.[31]

Yet were zero population growth to be achieved, so the thinking goes, the profligate use of resources would cease, or at least diminish. Advocates urge solving the population explosion through whatever means available, including U.N. programs. One problem with this approach is that U.N. programs are underfunded, and their effectiveness has been limited by so-called pro-life groups, especially those in the United States, who persuade the U.S. government to cut its funding for U.N. population-control efforts out of fear that the programs promote abortion. Another problem with this approach is that since population control is directed largely toward the nonwhite populations of Asia and Africa, it appears to have eugenic aims. In any event, the problem of how to slow population growth remains, and decisions to withhold funds from international family-planning agencies will have damaging consequences for resource-starved people who desire to limit the size of their families, consequences that often spill violently over borders.

Hardin also spoke of the need for coercion to bring about measured resource use. Clearly, some kind of coercive measures are required to prevent profligate use of resources: quotas, and taxes, fines, tariffs, and other economic disincentives, along with other measures to make wasteful use of resources politically and economically costly, in the absence of technological solutions. But aren't technological solutions in essence perpetual-motion machines? Given our growing population and finite resources, technology is no panacea. And yet, the ability of scientists and engineers to supply the citizens of the world with antibiotics, energy-efficient technologies, and agricultural produce for a healthy diet, suggests that quality of life can improve at the same time that we protect the natural environment.

Perhaps some technologies that are inappropriate in one setting may be appropriate in another. What, for example, is the value of bovine growth hormone (BGH), which will increase milk production in cattle 10 to 15 percent in a country such as the United States that already produces surplus amounts of milk products? The U.S. government spends millions of dollars annually to buy and store dairy products, at great expense to the consumer. The justification given is that the purchases keep small family farmers in business. (Bovine growth hormone, it turns out, has been adopted by agribusinesses, not small family farmers, who are disappearing from the nation's farmlands.) But in Russia or Zambia or another country with difficulty producing adequate quantities of milk, BGH might make possible a better diet for small children.

In 1972 four authors presented a deeply pessimistic view of the state of the earth in the now famous *Limits to Growth*. They argued that the problems of population growth, exhaustion of nonrenewable resources, rising malnutrition, and deterioration of the environment reflected the need for large-scale socioeconomic and political changes. Industrial economies whose success was based on constant growth could not be sustained. Equilibrium and stability in the economy and the environment were the only hope for the future.[32]

Whether the suggestions set forth in *Limits to Growth* were feasible or not, they raised important issues. The authors urged changes in domestic behavior, even more than changes in international behavior, to reduce consumption, fight pollution, and slow environmental degradation. International treaties and conventions had already had some success in promoting stewardship of living resources, they believed, but were successful only when the biggest offenders determined to participate. The absence of any technological cure-all, especially given the unwillingness of nations to change their pattern of resource use, suggested to the authors that optimism about the future of the earth's ecosystems was misplaced.

Treaty agreements, international organizations, and aid to impoverished nations are likely to offer some solutions to rapacious resource use and environmental degradation. This will be true particularly if the notion of human rights is made central to dealing with global environmental issues. This step requires a change in worldview, even though human rights issues have played a prominent role in international politics since the end of World War II and the Nuremberg trials. Another impetus to human rights was the effort by the U.S. government to use rights issues to isolate the USSR. By 1977, participants in the United Nations Conference on Desertification had declared that all people have the right to drinking water in quantities and of a quality to meet their basic needs. But only in the 1990s did specialists in the international arena acknowledge that environmental issues are connected with human rights. In May 1994, an international group of experts on human rights and environmental protection convened at the United Nations in Geneva and drafted the first-ever declaration of principles concerning human rights and the environment. According to this declaration, now embraced by the Office of the U.N. High Commissioner for Human Rights and the executive director for the U.N. Environment Programme, human rights cannot be secured in a degraded or polluted environment, and the fundamental right to life is threatened by soil degradation, deforestation, exposure to pesticides, chemicals, hazardous waste, and polluted water. According to the 1994 declaration, "human rights, an ecologically sound environment, sustainable development and peace are interdependent and indivisible," and "all persons have the right to a secure, healthy and ecologically sound environment. This right and other human rights, including civil, cultural, economic, political and social rights, are universal, interdependent and indivisible." The declaration further urged nations not to fob off on other people difficult resource management decisions: "All persons have the right to an environment adequate to meet equitably the needs of present generations and that does not impair the rights of future generations to meet equitably their needs." The declaration covered the export of dangerous and polluting industries to developing nations in its caveat that all persons have the right to a safe and healthy working environment, and "to adequate housing, land tenure and living conditions in a secure, healthy and ecologically sound environment."[33]

The example of water is revealing. Water quality is one of the most important human rights issues, not to mention a major public health issue. Many communities have access to ample amounts of clean water at low rates, while others pay exorbitant amounts for clean water, if they can get it at all. But the majority of people in the world have neither easy access to clean, safe water nor much of it. Almost half of Africans have no access to clean running water. Some three hundred million people on the continent also have no access to sanitation facilities, a situation that contributes to the high risk of contracting such infectious diseases as cholera and dysentery. In India more than a quarter of the population live in abject poverty; those states which have higher levels of groundwater have lower levels of poverty. Punjab, with only 6.16 percent of its people living below the poverty line, utilizes nearly all its groundwater resources.[34]

Another reason human rights considerations should play a role in the environmental movement is that in several cases governments have imprisoned environmental protesters, in clear violation of some of these principles. One such case concerns Alexander Nikitin, a former officer in the Soviet navy, who shared with Norwegian specialists in the Bellona Foundation nonclassified documents about the Soviet practice of dumping highly radioactive materials into the Arctic Ocean. The Russian government imprisoned Nikitin on three occasions and prosecuted him several times for trafficking in state secrets, before it finally exonerated him in 1998.

Ultimately, it is more than an empty slogan to suggest we ought to "think globally" and "act locally." Many of the world's environmental problems cross national boundaries: greenhouse gases, acid rain, pollution that enters rivers and then flows through several nations, and hazardous waste dumped on land or at sea in containers that eventually leak. Many problems are based on the high levels of consumption of nonrenewable resources. Those levels cannot be sustained. Further, other nations aspire to increase their levels of consumption and comfort, and as they do, they will put more pressure on resources. Are the only solutions to environmental problems in the twenty-first century international in character—for example, bilateral and multilateral treaties that coerce us to produce more efficiently and to approach consumption with other generations and people in mind? Why would any nation agree to an international arrangement to compel its citizens to do something that it will not commit to at a national level? The solutions will not be technological, but political and economic, and they will require a change in behavior and worldview, for the view of nature as machine that humans can perfect has brought us to the brink of disaster.

How paradoxical it is that in the twenty-first century, technology is simultaneously a solution for difficult environmental problems and a contributor to them, when employed without careful weighing of the uncertain social and environmental costs or the danger of technological failure. I hope that readers will continue to explore the many paradoxes of the interplay of the state, technology, and environment raised in this book, and I am confident that informed citizens will arrive at good solutions to persistent environmental problems.

Introduction

Mohandas K. Gandhi, from Pyarelal, *Towards New Horizons* p. 12, as cited in Vandana Shiva, "Ecology Movements in India," *Alternatives* 11 (1986): 255–273.

1. Costa Rica established its national parks system in 1970. Twelve percent of the country is protected as national parks, and another 16 percent as Indian reserves, biological reserves, wildlife refuges, and wildlife corridors—in all, more than a quarter of the land of Costa Rica. See *http://central america.com/cr/parks*.

2. See Richard White, "American Environmental History: The Development of a New Historical Field," *Pacific Historical Review* (August 1985): 297–335. For new perspectives, see William Cronon, editor, *Uncommon Ground: Rethinking the Human Place in Nature* (New York: W. W. Norton, 1996). The journal *Environmental History,* a publication of the American Society for Environmental History, has been published for more than twenty-five years.

3. Henry Wyes, "Hazardous Waste: Its Impact on Human Health in Europe," *Toxicology and Industrial Health* 13, nos. 2–3 (1997): 113–115.

4. Friedrich Klemm, *A History of Western Technology* (Cambridge: MIT Press, 1964).

5. Jacques Loeb, *The Mechanistic Conception of Life,* edited by Donald Fleming (Cambridge: Harvard University Press, 1964), 33.

6. Carolyn Merchant, *The Death of Nature: Women, Ecology and the Scientific Revolution* (San Francisco: Harper and Row, 1980).

7. Translated from Marie-Jean-Antoine-Nicholas Caritat, marquis de Condorcet, *Esquisse d'un tableau historique des progrès de l'esprit humain* (Paris: Masson et Fils, 1822), 279–285, 293–294, 303–305, at *http://www.fordham .edu/halsall/mod/condorcet-progress.html*. See also *http://ishi.lib.berkeley*

.edu/~hist280/research/condorcet/pages/progress_main.html.

8. David Noble, *American by Design: Science, Technology and the Rise of Corporate Capitalism* (New York: Oxford University Press, 1977); Jerome Ravetz, *Scientific Knowledge and Its Social Problems* (New York: Oxford University Press, 1971).

9. Jim Hightower, *Hard Tomatoes, Hard Times* (Rochester, VT: Schenkman Books, 1978).

10. Philip Kelley, "Blue Revolution or Red Herring? Fish Farming and Development Discourse in the Philippines," *Asia Pacific Viewpoint* 37, no. 1 (April 1996): 39.

11. Justus Laichena and James Wafula, "Biogas Technology for Rural Households in Kenya," *OPEC Review* 21 (1997): 223–244.

12. P. D. Dunn, *Appropriate Technology* (New York: Schocken Books, 1978), 3–5.

13. Shiva, "Ecology Movements."

14. D. G. Tendulkar, *Mahatma* (New Delhi: Ministry of Information and Broadcasting, 1953), 7: 35, 62–63.

15. M. K. Gandhi, *Hind Swaraj* (1938), 61, as cited in Shiva, "Ecology Movements."

16. Konrad Schliephake, "Irrigation and Food Production: Experiences from North Africa and Application to East Africa," *Applied Geography and Development* (Tübingen: Institute for Scientific Cooperation) 30: 30–45.

17. See William Cronon, *Changes in the Land* (New York: Hill and Wang, 1983), for discussion of the relative importance of capitalism, land-use practices among Native Americans and settlers, population pressures, and the like, on the changing ecology of New England. See also James Malin, "Man, the State of Nature, and Climax," *Scientific Monthly* 74, no. 1 (January 1952): 29–37.

1. The Modern State, Industry, and the Transformation of Nature

1. Quoted in L. S. Mulvin and P. Dillon Malone, "Air Pollution, Stone Decay and Irish Law," *Administration* 39, no. 4 (1992): 326.

2. Henry David Thoreau, *The Maine Woods* (New York: Bramhall House, 1950), 59–60, 67–68, 325–326.

3. George Perkins Marsh, *Man and Nature* (New York: Scribner, 1864).

4. See, for example, John Muir, *A Thousand-Mile Walk to the Gulf* (Boston: Houghton Mifflin, 1916) and Thoreau, *The Maine Woods.*

5. Samuel Hays, *Conservation and the Gospel of Efficiency* (Cambridge: Harvard University Press, 1959), 1–4.

6. Gifford Pinchot, *Breaking New Ground* (New York: Harcourt, Brace, 1947), 306–313.

7. James Scott, *Seeing Like a State* (New Haven: Yale University Press, 1998).

8. Pinchot, *Breaking New Ground,* 319–334. See also Char Miller, *Gifford Pinchot and the Making of Modern Environmentalism* (Washington, DC: Island Press, 2001).

9. For the bureau's history of itself, see *http://www.usbr.gov/history/borhist-.htm.*

10. Pinchot, *Breaking New Ground,* 335–339.

11. Ibid., 355.

12. *http://www.cr.nps.gov/local-law/anti-1906.htm.* Congressmen, primarily from Western states, seeing individual property rights as more important than any notion of public good have introduced legislation on several occasions to limit the presidential authority set forth in the Antiquities Act, notably in the 106th and 107th Congresses.

13. *http://www.yellowstone-online.com/history/NPS1916.html.* On "roads and motorized recreation on America's public lands," including the impact of snowmobiles and all-terrain vehicles (ATVs), see David Havlick, *No Place Distant* (Washington, DC: Island Press, 2002). In 2002, extensive debate arose over this power and the intent of the law because the administration of President George W. Bush cited the language to support the sale of timber, by claiming that it played a role in fire prevention.

14. On the early history of the Army Corps of Engineers, see Todd Shallat, *Structures in the Stream* (Austin: University of Texas Press, 1994).

15. Pete Daniel, *Deep'n as It Come: The 1927 Mississippi River Flood* (New York: Oxford University Press, 1977).

16. On the St. Paul flood control project, see *http://www.mvp.usace.army.mil/finder/display.asp?pageid=41.*

17. See "Wetlands and Flood Control in the Mississippi Watershed" at *http://www.crcwater.org/issues2/mississippi.html.* During the 108th Congress in 2003–4 there was an effort to release a third of the remaining wetlands for development. This move gives some idea of the difficulties in balancing economic, conservation, preservation, and other interests.

18. On the 1927 Mississippi flood, see Daniel, *Deep'n as It Come.*

19. The 1927 flood led to the 1928 flood-control act and an appropriation of

$300 million to the Army Corps of Engineers. During 1930s the corps, again turning to navigation, sought to deepen channels and constructed twenty-six locks and dams on the upper Mississippi alone. Since 1960 the corps has spent over $30 billion on flood-control efforts, yet flood damage has increased steadily to $3.5 billion annually. See "Wetlands and Flood Control in the Mississippi Watershed" at *http://www.crcwater.org/issues2/mississippi.html*.

20. On the massive projects of the Army Corps of Engineers and other organizations, see Marc Reisner, *Cadillac Desert: The American West and Its Disappearing Water* (New York: Penguin Books, 1986).

21. Studies that focus or touch on the Columbia River basin include Anthony Netboy, *The Salmon: Their Fight for Survival* (Boston: Houghton Mifflin, 1974), Blaine Harden, *A River Lost: The Life and Death of the Columbia* (New York: Norton, 1996), Murray Morgan, *The Columbia, Powerhouse of the West* (Seattle: Superior Publishing, 1949), and especially Richard White, *The Organic Machine* (New York: Hill and Wang, 1995).

22. Paul Josephson, *Industrialized Nature* (Washington, DC: Island Press, 2002), 17–62.

23. Donald Worster, *Dust Bowl* (New York: Oxford University Press, 1979).

24. Boyce Richardson, *James Bay: The Plot to Drown the North Woods* (San Francisco: Sierra Club, 1972).

25. Helmut Maier, *Elektrizitätswirtschaft zwischen Umwelt, Technik und Politik: Aspekte aus 100 Jahren RWE-Geschichte, 1898–1998* (Freiberg: Technische Universität Bergakademie, 1997).

26. Communication from Professor Thomas Zeller, October 28, 2002. On the RWE, see Maier, *Elektrizitätswirtschaft*.

27. Pinchot, *Breaking New Ground*, 319–334.

28. For highlights of Tennessee Valley Authority history, see *http://www.tva.gov/abouttva/abc/index.htm*.

29. David Lilienthal, *TVA: Democracy on the March* (New York: Harpers, 1953), 77.

30. S. Robert Aiken and Colin H. Leigh, "Hydro-Electric Power and Wilderness Protection," *Impact of Science on Society* no. 141: 85–96.

31. George Bennett, *Management of Lakes and Ponds*, 2d edition (Malabar, FL: Krieger, 1992), 26–27.

32. On the social and environmental aspects of aquaculture in different national settings, see Conner Bailey, Svein Jentoft, and Peter Sinclair, editors,

Aquacultural Development: Social Dimensions of an Emerging Industry (Boulder, CO: Westview Press, 1996).

33. Bennett, *Management of Lakes and Ponds*, 1–20.

34. Ibid., 65–68.

35. Netboy, *The Salmon*.

36. J. E. Thorpe, "The Development of Salmon Culture towards Ranching," in Thorpe, editor, *Salmon Ranching* (London: Academic Press, 1980) 1–11. See also J. T. Bowen, "A History of Fish Culture," in N. G. Benson, *A Century of Fisheries in North America* (Washington, DC: American Fish Society Special Publication 7, 1970).

There are other kinds of aquaculture. One is catfish aquaculture, which began in 1920s in warm-water Southern states that bordered the Mississippi River. On catfish aquaculture, see Karni Perez, Conner Bailey, and Amy Waren, "Catfish in the Farming System of West Alabama," in Bailey, Jentoft, and Sinclair, *Aquacultural Development*, 125–142.

Another is eel culture, which commenced in Japan in 1894. This farming of several species in high demand is energy-inefficient. By the mid-1970s, the industry produced twenty-four thousand metric tons annually, and the method had spread to Korea and Taiwan. Europe, New Zealand, and Australia are also interested in the techniques. This expansion occurred despite the fact that "in a protein-hungry world it is highly wasteful to culture carnivorous animals such as eels, but that is the way it goes." See Atsushi Usui, *Eel Culture* (Farnham, England: Fishing News Books, 1974). According to this leading Japanese eel culturalist, "When a Dutchman eats an eel he likes to feel the oil trickling out of the corners of his mouth and down his chin. When a German eats an eel he likes to bite something big and solid."

37. William Warner, *Distant Water* (Boston: Little, Brown, 1983).

38. On Soviet practices, see Vladil Lysenko, *A Crime against the World*, translated by Michael Glenny (London: Victor Gollancz, 1983).

39. In *Common Lands, Common People* (Cambridge: Harvard University Press, 1997), Richard Judd discusses the politics of disputes over community-based versus scientifically determined knowledge.

40. Tim Smith, *Scaling Fisheries* (Cambridge: Cambridge University Press, 1994).

41. Miriam Wright, *A Fishery for Modern Times* (New York: Oxford University Press, 2001), 1–6.

42. Ibid., 10–41.

43. Ibid.

44. Cato Wadel, "Capitalization and Ownership: The Persistence of Fishermen-Ownership in the Norwegian Herring Industry," *Newfoundland Social and Economic Papers* 5 (1972): 116–118.

45. Ole Strandgaard, "Salmon Fishing and/or Animal Husbandry," *Folk* 16–17 (1974–1975): 233–242.

46. "The Norwegian Fishing Industry," at *http://odin.dep.no/fid/engelsk.*

47. Svein Jentoft and Knut Mikalsen, "Regulating Fjord Fisheries: Folk Management or Interest Group Politics?" in Christopher Dyer and James McGoodwin, editors, *Folk Management in the World's Fisheries* (Niwot: University of Colorado Press, 1994), 287–288.

48. Andrew Kimbrell, *The Fatal Harvest Reader: The Tragedy of Industrial Agriculture* (Washington, DC: Island Press, 2002).

49. Ralph D. Christy and Lionel Williamson, editors, *A Century of Service: Land-Grant Colleges and Universities, 1890–1990* (New Brunswick, NJ: Transaction Publishers, 1992).

50. Jim Hightower, *Hard Tomatoes, Hard Times* (Rochester, VT: Schenkman Books, 1973).

51. Leland Swenson, "Sowing Disaster," *Forum for Applied Research and Public Policy* 14, no. 3 (Fall 1999): 48–54.

52. Ibid.

53. "The Rap Sheet on Animal Factories" at *http://www.sierraclub.org/factory farms.*

54. See Eric Schlosser, *Fast Food Nation* (New York: Perennial, 2002), for an indictment of the industrial approach to restaurant management in postwar America. The trend toward concentration has been exacerbated by legislation at the beginning of the twenty-first century. The Farm Security and Rural Investment Act of 2002 authorized $248.6 billion in spending that increases taxpayer obligations for agriculture by more than 80 percent over the 1996 farm bill, itself a failed attempt to free farmers from price supports and commodity payments, as the United States was bound to do under its World Trade Organization (WTO) obligations. The top tenth of farm-subsidy recipients will collect two-thirds of the money, and the bottom four-fifths get just one-sixth. Almost half of commodity payments will go to large farms with average household incomes of $135,000. See Anuradha Mittal, "Giving Away the Farm: The 2002 Farm Bill," at *http://www.foodfirst.org/pubs/backgrdrs/2002/s02v8n3.html.*

55. On the history of genetically modified organisms and the food industry, see Daniel Charles, *Lords of the Harvest* (Cambridge: Perseus, 2001). On the

ethical, social, and other dimensions of genetic engineering for humans, and on the troubling relationship between corporations and universities supported by public funds whose research benefits corporations, see Sheldon Krimsky, *Biotechnics and Society* (New York: Praeger, 1991).

56. On the environmental costs of Hanford Atomic Reservation, as only one example of the hundreds of billions of dollars it will require to control dangerous radioactive pollution, see Michael D'Antonio, *Atomic Harvest* (New York: Crown, 1993).

57. About dry aboveground spent-fuel storage casks, see *http://www.maine yankee.com.*

58. Over the years, scientists proposed storing waste underground in caverns created by "peaceful nuclear explosions" and even ferrying it to the sun in space transports that would burn up, together with their noxious cargo, in the sun's atmosphere. Of course, were a transport to explode in the earth's atmosphere as the space shuttles Challenger and Columbia did, the waste would spread far and wide.

59. On nuclear power in France, see Gabrielle Hecht, *The Radiance of France* (Cambridge: MIT Press, 1998). See also Bertrand Goldschmidt, *The Atomic Complex* (Le Grange Park, IL: American Nuclear Society, 1982).

60. Peter Faulkner, editor, *The Silent Bomb* (New York: Random House, 1977), 3–22, 43–63. Another issue affecting environmental safety is the fact that many nuclear power stations have reached the end of their lives prematurely. Engineers failed to understand the rapid aging of plants and their components under the influence of high temperature, pressure, and creep and corrosion caused by radioactivity. Reactor owners and operators, clamoring to prolong the life of the stations, maintain that the reactors can be run safely, even though as the electrical power industry has been deregulated. Can regulatory agencies balance the need to produce enough electricity to maintain the current standard of living against the need to shut down stations that have operated longer than intended?

61. Edward Charles Jeffrey, *The Origin and Organization of Coal* (Lancaster, PA: American Academy of Arts and Sciences, 1924) 15, no. 1: 5–6.

62. Mulvin and Malone, "Air Pollution, Stone Decay and Irish Law."

63. D. T. Randall, *The Burning of Coal without Smoke in Boiler Plants* (Washington, DC: USGPO, 1908), USGS bulletin no. 34, series Q, Fuels 8.

64. See James Fleming and Bethany Knorr, "History of the Clean Air Act," at *http://www.ametsoc.org/sloan/cleanair/.*

65. Krishna Dhir, Joann Stewart, and Willie Hopkins, "Coal: A Diminishing

Hope for America's Energy Needs," *Business and Society* 22 (Spring 1983): 35–39.

66. Paul G. Rogers, "The Clean Air Act of 1970," in the *EPA Journal* (January–February 1990), as posted at *http://www.epa.gov/history/topics/caa70/11 .htm*. In 2003, the administration of George W. Bush announced its "Clean Skies" program, which will enable fossil fuel plants to avoid modernizing and thereby to pollute more heavily.

67. Mulvin and Malone, "Air Pollution, Stone Decay and Irish Law."

68. "Love Canal" at *http://www.epa.gov/region02/superfund/npl/0201290c.htm*.

69. Lawrence Fishbein, "Municipal and Hazardous Waste Management: An Overview," *Toxicology and Industrial Health* 7, nos. 5–6 (1991): 209–218. See also OTA, *Serious Reduction of Hazardous Wastes* (Washington, DC: USGPO, 1986), and OTA, *From Pollution to Prevention* (Washington, DC: USGPO, 1987).

70. Craig Colten, "Historical Questions in Hazardous Waste Management," *Public Historian* 10, no. 1 (Winter 1988): 7–20.

71. Fishbein, "Municipal and Hazardous Waste."

72. Colten, "Historical Questions in Hazardous Waste." The Bush administration also joined a failed effort by General Electric Corporation to overturn provisions of CERCLA that required it to clean up carcinogenic PCBs from the Hudson River in upstate New York. See "Bush Administration Says It Won't Seek Superfund Tax Reauthorization," at *http://www.publicagenda .org/headlines/022502headline.htm*.

73. On the "garbage crisis" in America, see Martin Melosi, *Effluent America* (Pittsburgh: University of Pittsburgh Press, 2001), 68–91.

74. "Fresh Kills: You Can't Fill a Bottomless Pit," at *http://www.johnmccrory .com/bags/history/history2.html*, and "Landfill into Landscape," at *http:// www.nyc.gov/html/dcp/html/fkl/ada/about/1_2.html*.

75. Robert Bullard, *Dumping in Dixie: Race, Class and Environmental Quality* (Boulder, CO: Westview Press, 1994).

76. Evan Ringquist, "A Question of Justice: Equity in Environmental Litigation, 1974–1991," *Journal of Politics* no. 4 (November 1998): 1148–1165. Yet on the basis of his investigation of the enforcement of the Clean Air Act, Clean Water Act, and Superfund (RCRA), Ringquist concluded there is "no significant relationship between race, income, and total civil fines."

77. Adam Rome, *The Bulldozer in the Countryside: Suburban Sprawl and the Rise of American Environmentalism* (New York: Cambridge University Press, 2001).

78. "Wastewater Management: Alternative Small-Scale Treatment Systems," *Management Information Service* 17 (April 1985), 1–23. Sometimes facilities must add chemicals to cause such dangerous substances as phosphorus and nitrogen to precipitate out or join with other compounds to lessen their toxicity. Aerated lagoons or stabilization ponds are a low-cost alternative to conventional systems, but they require more time and space, if less energy and personnel.

79. Rome, *The Bulldozer in the Countryside.*

80. Paul Ruffins and Marsha Coleman-Adebayo, "Poverty, Politics and Environmental Pollution in Africa," *Focus* 20 (January 1992): 5–6.

81. Seid M. Zekavat, "The State of the Environment in Iran," *Journal of Developing Societies* 13, no. 1 (1997): 49–72.

82. Iran also turned to heavy use of chemical fertilizers, pesticides, and herbicides in agriculture, with 605,000 kilograms of poisons used for weed elimination and three groups of pests in Fars Province in 1992 alone, and an increase in the use of fertilizers from 5,000 tons to 1.7 million tons in the second half of the twentieth century, yet with no significant increase in the agricultural land under cultivation. The result was "a declining net national welfare." See ibid.

83. For a history of the development of the SUV and its outrageous environmental and other public health costs, see Keith Bradsher, *High and Mighty* (New York: Public Affairs, 2002).

84. Raymond H. Dominick III, *The Environmental Movement in Germany* (Bloomington: Indiana University Press, 1992), 43–57.

85. Ibid., 81–85.

86. Raymond Dominick, "The Roots of the Green Movement in the United States and West Germany," *Environmental Review* 12, no. 3 (1988): 1–17.

87. Ibid., 20–23.

88. Rachel Carson, *Silent Spring* (New York: Fawcett Crest, 1962), 24–43.

89. Donella Meadows, Dennis Meadows, Jorge Randers, and William Behrens, *Limits to Growth: A Report for the Club of Rome's Project on the Predicament of Mankind* (New York: Universe Books, 1974); Dominick, "The Roots of the Green Movement," 1–17.

90. Rolf Lidskog and Ylva Uggla, "Mercury Waste Management in Sweden," *Journal of Environmental Planning and Management* no. 4 (July 2000): 561–570.

91. Peter Michaelis, "Product Stewardship, Waste Minimization and Economic Efficiency: Lessons from Germany," *Journal of Environmental Planning and*

Management 38, no. 2 (June 1995): 231–244.

92. Ibid.

93. Lisa Heinzerling, "Regulatory Costs of Mythic Proportions," *Yale Law Journal* 107 (1998): 1981–2070.

94. Carson, *Silent Spring,* 261.

2. The Coercive Appeal to Order

1. Douglas Weiner, *Models of Nature* (Bloomington: Indiana University Press, 1988), and *A Little Corner of Freedom* (Berkeley: University of California Press, 1999).

2. Marshall Goldman, *Spoils of Progress* (Cambridge: MIT Press, 1972).

3. Murray Feshbach and Alfred Friendly, Jr., *Ecocide* (New York: Basic Books, 1992).

4. Jeffrey Stine, *Mixing the Waters* (Akron, OH: University of Akron Press, 1993), 54–57.

5. Michael Bressler, "Agenda Setting and the Development of Soviet Water Resources Policy, 1965–1990," Ph.D. dissertation, Department of Political Science, University of Michigan, 1992, and G. V. Voropaev and D. Ia. Ratkovich, *Problema territorial'nogo pereraspredeleniia vodynkh resursov* (The problem of territorial distribution of water resources) (Moscow: IVP AN SSSR, 1985).

6. Zhores Medvedev, *Nuclear Disaster in the Urals* (New York: W. W. Norton, 1979).

7. For a brief and comprehensive discussion of the extensive cumulative impact of these radiation disasters on the Cheliabinsk region, see Scott Monroe, "Cheliabinsk: The Evolution of Disaster," *Post-Soviet Geography* 33 (1992): 533–545. See also Paul Josephson, *Red Atom* (New York: W. H. Freeman 1999), 279–280.

8. Josephson, *Red Atom.*

9. Ibid.

10. Yulian Konstantinov, "The 'Dragon of Kovachitsa': Local Perceptions of Radioactive Pollution near the Kozlodui Nuclear Power Station (Bulgaria)," *Human Ecology* 23, no. 1 (1995): 99–110.

11. Robert Darst, *Smokestack Diplomacy* (Cambridge: MIT Press, 2001), 147–148.

12. Jane Dawson, *Econationalism* (Durham, NC: Duke University Press, 1996).

13. Liliana B. Andonova, "The Challenges and Opportunities for Reforming Bulgaria's Energy Sector," *Environment* 44, no. 10 (December 2002): 9–19.

14. Jeffrey Herf, *Reactionary Modernism: Technology, Culture and Politics in Weimar and the Third Reich* (Cambridge: Cambridge University Press, 1984).

15. Raymond H. Dominick III, *The Environmental Movement in Germany* (Bloomington: Indiana University Press, 1992), 85–96. Quotation from Verein Naturschutzpark, *Der erste deutsche Naturschutzpark in der Lüneburger Heide* (Stuttgart: Frank'sche Verlagsbuchhandlung, 1920), 9, as cited in Dominick, *Environmental Movement,* 88.

16. Dominick, *Environmental Movement,* 96–106.

17. Ibid., 111–114.

18. Karl Ditt, "The Perception and Conservation of Nature in the Third Reich," *Planning Perspectives* 15, no. 2 (2000): 161–187, and "Nature Conservation in England and Germany, 1900–1970: Forerunner of Environmental Protection?" *Contemporary European History* 5, no. 1 (1996): 1–28.

19. Thomas Zeller, "Landschaften des Verkehrs: Autobahnen im Nationalsozialismus und Hochgeschwindigkeitsstrecken für die Bahn in der Bundesrepublik," *Technikgeschichte* 64 (1997): 323–340.

20. Dominick, *Environmental Movement,* 106–111.

21. Mechtild Rossler, "'Area Research' and 'Spatial Planning' from the Weimar Republic to the German Federal Republic: Creating a Society with a Spatial Order under National Socialism," in Monika Renneberg and Mark Walker, *Science, Technology and National Socialism* (New York: Cambridge University Press, 1994), 126–138.

22. Judith Shapiro, *Mao's War against Nature: Politics and Environment in Revolutionary China* (Cambridge: Cambridge University Press, 2001).

23. Vaclav Smil, *The Bad Earth: Environmental Degradation in China* (Armonk, NY: M. E. Sharpe, 1984), 80–81.

24. Ibid., 82–93.

25. Ibid., 45–47.

26. Wu Ming, "Major Problems Found in Three Gorges Dam Resettlement," *China Rights Forum* (Spring 1998): 4–9.

27. Sukhan Jackson and Adrian Sleigh, "Resettlement for China's Three Gorges Dam: Socio-Economic Impact and Institutional Tensions," *Communist and Post-Communist Studies* 33 (2000): 223–241.

28. Ibid.

29. Wu Ming, "Major Problems," 4–9.

30. Shiu-Hung Luk and Joseph Whitney, editors, *Megaproject: A Case Study of China's Three Gorges Project* (Armonk, NY: M. E. Sharpe, 1993), 3–39.

31. Wu Mei, "Uncovering Three Gorges Dam," *Media Studies Journal* 13 (Winter 1999): 122–129.

32. Wu Ming, "Major Problems," 4–9.

33. Jackson and Sleigh, "Resettlement for China's Three Gorges Dam," 223–241.

34. Robert Wirtshafter and Ed Shih, "Decentralization of China's Energy Sector: Is Small Beautiful?" *World Development* 18, no. 4 (1990): 505–512.

35. Huming Yu, "China's Coastal Ocean Uses: Conflicts and Impacts," *Ocean and Coastal Management* 23 (1994): 161–178.

36. Ibid.

37. Fudao Zhang, et al., "Urban Solid Waste in China: Current Problems and Solutions," *Chinese Geography and Environment* 2 (Spring 1989): 54–67; and Smil, *The Bad Earth*, 100–118.

38. Hong Cheng, "The Situation and Prospect of Forestry in China," *Land Use Policy* 6 (January 1989): 64–74.

39. Smil, *The Bad Earth*, 9–62.

40. Ibid., 127–135, 171.

41. Sergio Osmena, *The Commonwealth: A Year of Accomplishments, 1938–39* (Manila: Bureau of Printing, 1941), 7–8.

42. Ferdinand Marcos, *Presidential Speeches* (n.p.: 1978), vols. 1–6.

43. See *http://www.antenna.nl/wise/397/3863.html*.

44. Ferdinand E. Marcos, *Five Years of the New Society* (Manila, 1978), 29, 48, 57, 71–77; and Marcos, *Essays in Aspects of Philippine Development toward the New Society* (Manila: National Media Production Center, 1974).

45. Philip Kelley, "Blue Revolution or Red Herring? Fish Farming and Development Discourse in the Philippines," *Asia Pacific Viewpoint* 37, no. 1 (April 1996): 39–57.

46. Ibid., 39–57.

47. Golbery do Couto e Silva, *Aspectos Geopoliticos do Brasil* (Rio de Janeiro: Biblioteca do Exercito, 1957), and Susanna B. Hecht, "Cattle Ranching in Amazonia: Political and Ecological Considerations," in Marianne Schmink and Charles H. Wood, editors, *Frontier Expansion in Amazonia* (Gainesville, FL: University of Florida Press, 1984), 367–370.

48. Paul Josephson, *Industrialized Nature* (Washington, DC: Island Press, 2002), 142.

49. Susanna Hecht and Alexander Cockburn, *The Fate of the Forest* (London: Verso, 1989), 100.

50. Emilio F. Moran, "Colonization in the Transamazon and Rondonia," in Schmink and Wood, *Frontier Expansion in Amazonia,* 292–297; and Hecht, "Cattle Ranching in Amazonia," 374–377.

51. John O. Browder, "Public Policy and Deforestation in the Brazilian Amazon," in Robert Repetto and Malcolm Gillis, editors, *Public Policies and the Misuse of Forest Resources* (Cambridge: Cambridge University Press, 1988), 255; Moran, "Colonization in the Transamazon and Rondonia," 287–291; Donald R. Sawyer, "Frontier Expansion and Retraction in Brazil," in Schmink and Wood, *Frontier Expansion in Amazonia,* 188–189; Emilio F. Moran, "Government-Directed Settlement in the 1970s: An Assessment of Transamazon Highway Colonization," in Emilio F. Moran, *The Dilema of Amazonian Development* (Boulder, CO: Westview Press, 1983), 302–309; Binka Le Breton, *Voices from the Amazon* (West Hartford, CT: Kumarian Press, 1993), 73; and A. U. Oliveira, *Amazônia: Monopólio, expropriação e conflitos,* 2d ed. (São Paulo: Papirus, 1989).

52. Marianne Schmink, "Big Business in the Amazon," in Julie Denslow and Christine Padoch, editors, *People of the Tropical Rain Forest* (Berkeley: University of California Press, 1988), 172–173; and Le Breton, *Voices from the Amazon,* 42–55.

53. Tarciana Portella, *Itaparica: A Dor de um Povo Gerando Energia* (The hydroelectric plant at Itaparica: A struggle with power), translated by Sarah Bailey (Recife, Brazil: Centro de Defesa dos Direitos Humanos do Submedio São Francisco, 1992), 16; and Leinad Ayer de O. Santos, Lucia M. M. de Andrade, editors, *Hydroelectric Dams on Brazil's Xingu River and Indigenous Peoples,* translated by Robin Wright (Cambridge, MA: Cultural Survival, 1990), 20–23.

54. Sebastião Pinheiro, *Tucurui: O Agente Laranja em uma República de Bananas* (Porto Alegre, Brazil: Sulina, 1989); http://www.dams.org/studies/br/; and Juan de Onis, *The Green Cathedral* (New York: Oxford University Press, 1992), 117–128. The Amazon aluminum industry, with investments of $21.5 billion, has created six thousand jobs, but the industry is highly sensitive to international aluminum prices.

55. Oliver Greenwood, "The Yvypyte Wood Scandal in Paraguay: A Case of Ecocide," *Survival International* 3, no. 14 (1978): 4–5.

3. Development, Colonialism, and the Environment

1. Stephen Lawani, "Africa's Quiet Revolution," *Standard Chartered Review* (October 1986): pp. 2–7.

2. Paul Ruffins and Marsha Coleman-Adebayo, "Poverty, Politics and Environmental Pollution in Africa," *Focus* 20 (January 1992): 5–6.

3. Ibid.

4. Bill Rau, *From Feast to Famine* (London: Zed Books, 1991), 146–150.

5. Ibid., 9–14.

6. Ibid., 14–19.

7. John Sanders, Barry Shapiro, and Sunder Ramaswamy, *The Economics of Agricultural Technology in Semiarid Sub-Saharan Africa* (Baltimore: Johns Hopkins University Press, 1996), 28–38.

8. Sara Berry, *No Condition Is Permanent* (Madison: University of Wisconsin Press, 1993), 3–22.

9. Melissa Leach and Robin Mearns, editors, *The Lie of the Land* (London: Villiers Publications, 1996), 16.

10. William Beinart, "Soil Erosion, Conservationism and Ideas about Development: A Southern African Exploration, 1900–1960," *Journal of Southern African Studies* 11, no. 1 (October 1984): 52–83. See also David Anderson, "Depression, Dustbowl, Demography and Drought: The Colonial State and Soil Conservation in East Africa during the 1930s," *African Affairs* 83, no. 332 (July 1984): 321–343.

11. Rau, *From Feast to Famine*, 146–150.

12. Annual Report of the Agricultural Department as cited in Mary Tiffen, Michael Mortimore, and Francis Gichuki, *More People, Less Erosion: Environmental Recovery in Kenya* (Chichester, England: Wiley, 1994), 252.

13. Leach and Mearns, *Lie of the Land*, 1–2.

14. Ibid., 1–2.

15. Ibid., 19–20.

16. Ibid., 2–3.

17. James Scott, *Seeing Like a State* (New Haven, CT: 1998).

18. Leach and Mearns, *Lie of the Land*, 5–8. They note that connected with this naïveté is the habit of seeing people both as passive groups and as quantifiable rather than as active subjects with agendas of their own and the knowledge to carry them out. The authors conclude that the cultural dimensions of good-bad dichotomies and of seeing problems outside their

specific historical and geographical context "reflect the hegemony of Western development discourse."

19. Fantu Cheru, *The Silent Revolution in Africa* (London: Zed Books, 1989), 2–5.

20. Paul Harrison, *The Greening of Africa* (New York: Penguin, 1987), 27–28, 114–140.

21. Mary Tiffen, Michael Mortimore, and Francis Gichuki conducted a thorough study of the relation between population growth and environmental degradation that helps to dispel the "received wisdom" about the neo-Malthusian roots of crisis. Evaluating data and observations about the impact of population growth on the Machakos district in southeast Kenya over sixty years, the authors conclude that population increase is compatible with recovery from environmental degradation. Degradation stemmed from such pressures as colonial land practices to support European settlement that pushed indigenous people into confined regions. Tiffen, Mortimore, and Gichuki's research showed an increase in output per capita and per hectare as lands recovered from the loss of vegetation and erosion. The authors acknowledge displacement of flora and fauna by humans—for example, through expansion of farmland into "natural" vegetation. But humans must farm, and farmers recognize and help promote natural resilience in the environment. They will regularly deplete, and then restore or improve, fertility through good farming practices. See Ibid., 3–25.

22. Tiffen, Mortimore, and Gichuki provide extensive evidence of the reversal of environmental damage in the Machakos district, rising productivity and living standards, and successful exploitation of lands previously deemed unfit for agriculture. This evidence led them to the conclusion that population growth, in Machakos and other regions, was accompanied by specialization, economic diversification, rising living standards, and an increasing rate of technological change, "which has outpaced any threat to the depletion of resources" (Ibid.).

23. Paul Harrison, *The Greening of Africa* (New York: Penguin, 1987), 114–140.

24. Konrad Schliephake, "Irrigation and Food Production: Experiences from North Africa and Application to East Africa," *Applied Geography and Development* (Tübingen: Institute for Scientific Cooperation, n.d.) 30: 30–45.

25. Harrison, *The Greening of Africa*, 98–113, 141–170.

26. Sanders, Shapiro, and Ramaswamy, *Economics of Agricultural Technology*, 28–38; Harrison, *The Greening of Africa*, 17–18.

27. Bondi Ogolla, "Water Pollution Control in Africa: A Comparative Legal Survey," *Journal of African Law* 33 (Autumn 1989): 149–156.

28. Clevo Wilson, "Environmental and Human Costs of Commercial Agricultural Production in South Asia," *International Journal of Social Economics* 27, nos. 7, 8, 9, and 10 (2000): 817–818.

29. Stephen Lawani, "Africa's Quiet Revolution," *Standard Chartered Review* (October 1986): pp. 2–7.

30. Berry, *No Condition Is Permanent*, 181–183, 196–201.

31. B. K. Darkoh, "Towards Sustainable Development and Environmental Conservation in African Drylands," *Journal of Eastern African Research and Development* 23 (1993): 1–23.

32. Lawani, "Africa's Quiet Revolution," 2–7. On International Institute of Tropical Agriculture cassava research, see *http://www.iita.org/crop/cassava.htm*.

33. Tiffen, Mortimore, and Gichuki, *More People, Less Erosion*; Sanders, Shapiro, and Ramaswamy, *The Economics of Agricultural Technology*, 28–31.

34. Cheru, *The Silent Revolution*, 6–7.

35. Ibid., 72–100.

36. Susan Andreatta, "Agrochemical Exposure and Farmworker Health in the Caribbean: A Local/Global Perspective," *Human Organization* 57, no. 3 (1998): 350–358.

37. Ibid.

38. Wilson, "Environmental and Human Costs," 819–823.

39. Ibid., 824–826.

40. Ibid., 827–834.

41. Ibid., 835.

42. Frederick Buttel, Martin Kenney, and Jack Kloppenburg, Jr., "From Green Revolution to Biorevolution: Some Observations on the Changing Technological Bases of Economic Transformation in the Third World," *Economic Development and Cultual Change* 34 (October 1985): 31–55.

43. Cheru, *The Silent Revolution*, 2–5.

44. Darkoh, "Towards Sustainable Development and Environmental Conservation in African Drylands," 1–23.

45. Philip Howard, "Development-Induced Displacement in Haiti," *Refuge* 16, no. 3 (August 1997): 4–11.

46. Ibid.

47. Kole Ahmed Shettima, "Dam Politics in Northern Nigeria: The Case of the Kafin Zaki Dam," *Refuge* 16, no. 3 (August 1997): 18–22.

48. Ibid.

49. Ibid.

50. E. O. Adeniyi, "The Kainji Dam: An Exercise in Regional Planning," *Regional Studies* 10 (1976): 233–243.

51. E-mail communication from Elizabeth Bishop, American University of Cairo, January 15, 2003.

52. Richard Elliot Benedick, "The High Dam and the Transformation of the Nile," *Middle East Journal* 33 (Spring 1979): 119–144.

53. Ibid.

54. "The Building of the First Aswan Dam and the Inundation of Lower Nubia," at *http://www.umich.edu/~kelseydb/Exhibits/AncientNubia/PhotoIntro.html.*

55. Benedick, "The High Dam," 119–144.

56. Donald Heyneman, "Dams and Disease," *Human Nature* 2, no. 2 (December 1979): 50–57. Heyneman writes: "One of the tragedies of modern technology is that we are incapable of predicting or controlling its ecological impact." He continues by observing that no matter who introduces the innovation, it "sometimes seems to create two new problems for every one it solves."

57. Madhav Gadgil and Ramachandra Guha, *This Fissured Land,* in *The Use and Abuse of Nature* (New Delhi, India: Oxford University Press, 2000), 115–121.

58. Ibid., 122–125, 134–135, 138–139.

59. Ibid., 146–149, 158–159.

60. Ibid., 162–164, 171–173.

61. For further discussion of Gandhi's social philosophy in *Hind Swaraj* (1909) and his indictment of modern Western civilization and its industries for destroying "organic village communities," see Gadgil and Guha, *Ecology and Equity,* in *The Use and Abuse of Nature,* 181–183.

62. Ashok Parthasarathi, "Science and Technology in India's Search for a Sustainable and Equitable Future," *World Development* 18, no. 12 (1990): 1693.

63. Ibid., 1694.

64. Ibid. 1694–1695.

65. Fred Pearce, "Building Disaster: The Monumental Folly of India's Tehri Dam," *The Ecologist,* vol. 21, no. 3 (May/June 1991): 123–28.

66. Ibid.

67. Ibid. In another case, the frequency of flooding around the Hirakud Dam in Orissa, India, actually accelerated from every twelve to every four years,

and its reservoir submerged 285 villages. The residents received inadequate compensation for their land. See Balgovind Baboo, "Development and Displacement: A Comparative Study of Rehabilitation of Dam Oustees in Two Suburban Villages of Orissa," *Man in India* 72, no. 1 (1992): 1–14.

68. See *http://www.narmada.org/sardarsarovar.html.*

69. See *http://www.sardarsarovardam.org/default.htm.* On the fate of the residents ousted because of Indian state irrigation and hydroelectricity projects, especially the gargantuan Narmada River valley projects, see Gadgil and Guha, *Ecology and Equity,* 56–57 and 72–73.

70. Gregory Holbrook, "Water of Life: The Social Effect of an Irrigation Scheme," *South African Journal of Ethnology* 20, no. 1 (1997): 39–42.

71. Philip Hirsch, "Large Dams, Restructuring and Regional Integration in Southeast Asia," *Asia Pacific Viewpoint* 37, no. 1 (April 1996): 1–20.

72. Parthasarathi, "Science and Technology," 1698–1699.

73. Jennifer Clapp, "The Toxic Waste Trade with Less-Industrialised Countries," *Third World Quarterly* 15, no. 3 (September 1994): 505–519; and Ruffins and Coleman-Adebayo, "Poverty, Politics and Environmental Pollution in Africa," 5–6.

74. Clapp, "The Toxic Waste Trade."

75. John Ovink, "Transboundary Shipments of Toxic Waste," *Dickinson Journal of International Law* 13 (Winter 1995): 281–295.

76. Ibid.

77. Joshua Karliner, "Corporate Power and Ecological Crisis," *Global Dialogue* 1, no. 1 (Summer 1999): 124–138.

78. G. M. Iskander and Badr El Din Khalil, "Nuclear Waste Disposal: A Commentary, with Discussion of a Possible Site in Sudan," *Sudan Notes and Records* 64 (1983): 203–222.

79. Ruben Berrios, "Technological Dependence and Alternative Development Strategies," *Scandinavian Journal of Development Alternatives* 4 (March 1985): 121–139.

80. Karliner, "Corporate Power and Ecological Crisis."

81. Berrios, "Technological Dependence."

82. Ibid., 126.

83. Priya Kurian, "Banking on Gender: Uncovering Masculinism in the World Bank's Environmental Policies," *Asian Journal of Public Administration* 21, no. 1 (June 1999): 55–85.

84. Harrison, *The Greening of Africa,* 56.

85. Ibid., 56–57.

4. Biodiversity, Sustainability, and Technology

1. James Rodger Fleming, "Global Environmental Change and the History of Science," in Mary Jo Nye, editor, *The Cambridge History of Science: The Modern Physical and Mathematical Sciences*, vol. 5 (Cambridge: Cambridge University Press, 2003), 634–650. See also Fleming, *Historical Perspectives on Climate Change* (New York: Oxford University Press, 1998).

2. See *http://www.foe.org/WSSD/bushadministration.html*. President Bush himself dismissed as "a product of bureaucracy" a report of his own government that confirmed the danger of growing reliance on fossil fuel consumption.

3. Kenneth Weiss, "Summit Scoffs at U.S. Pitch on Fixing Global Ills," *Los Angeles Times* at *http://www.smh.com.au/articles/2002/08/30/1030508125099 .html*. Representatives of other governments and many U.S. citizens in Johannesburg were skeptical of the U.S. position. They consider the partnerships to be no more than a few pilot projects. What should be expected from the world's richest nation? Jeffrey Sachs, director of the Earth Institute at Columbia University, says, "Solving the problems of some villages is not an appropriate stance for the United States. It puts us at risk in our foreign policy. We cannot lead the world in a war against terrorism if we don't lead the world in the war on poverty, disease and environmental degradation" (at *http://www.smh.com.au/articles/2002/08/30/103050812509g .html*).

4. United Nations, *A Quiet Revolution: The United Nations Convention on the Law of the Sea* (New York: Department of Public Information, United Nations, 1984). See also Elisabeth Borgese, *The Future of the Oceans: A Report to the Club of Rome* (Montreal: Harvest House, 1986).

5. Ted McDorman, "International Fishery Relations in the Gulf of Thailand," *Contemporary Southeast Asia* 12, no. 1 (June 1990): 40–54. McDorman writes that a solution to the problem of Thai vessels overfishing in other nations' waters might be for Thai fishermen, in exchange for fees, technology transfer, or profit sharing, to obtain official access to fish in the waters of neighboring states.

6. Kal Raustiala and David G. Victor, "Conclusions," in Victor, Raustiala, and Eugene Skolnikoff, editors, *The Implementation and Effectiveness of International Environmental Commitments: Theory and Practice* (Cambridge: MIT Press, 1998), 659–708.

7. William Conway, "Can Technology Aid Species Preservation?" in E. O. Wil-

son, editor, *Biodiversity* (Washington: National Academy Press, 1988), 263–268.

8. E. O Wilson, "The Current State of Biodiversity," in Wilson, *Biodiversity* 4–5.

9. For a historical study of the intentional and unintentional consequences of the introduction of non-native species in the Americas, see Alfred Crosby, *Ecological Imperialism: The Biological Expansion of Europe, 900–1900* (Cambridge: Cambridge University Press, 1986).

10. See *http://darwin.bio.uci.edu/~sustain/h90/Brazil.htm*.

11. See *http://forests.org/archive/brazil/brcbiodb.htm*.

12. Julie M. Feinsilver, "Biodiversity Prospecting: A New Panacea for Development?" *Cepal Review* 60 (December 1996): 115–132.

13. Ibid.

14. See *http://www.nmfs.noaa.gov/prot_res/laws/ESA/ESA_home.html*. For a list of endangered animals in United States, see *http://endangered.fws.gov/50cfr_animals.pdf*. Some of the first mammal species to be listed in the United States included the Indiana bat, the Delmarva Peninsula fox squirrel, the timber wolf, the grizzly bear, and the Florida manatee; among birds the Hawaiian dark-rumped petrel, the Tule white-fronted goose, the California condor, the Southern bald eagle, and the whooping crane; among reptiles and amphibians the American alligator, the blunt-nosed leopard lizard, and the Texas blind salamander; and among fish the shortnose sturgeon, a number of trout, the Devils Hole pupfish and the blue pike. See *http://endangered.fws.gov/1966listing.html*.

15. See *http://www.nmfs.noaa.gov/prot_res/laws/ESA/ESA_home.html*. On-line versions of the application instructions for *Scientific Research or Enhancement Permits for Endangered and Threatened Species,* for *Incidental Take Permits for Endangered and Threatened Species,* and for *Incidental Take Permits for Sea Turtles* are available.

16. Jeffrey K. Stine, *Mixing the Waters: Environment, Politics and the Building of the Tennessee-Tombigbee Waterway* (Akron, OH: University of Akron Press, 1994).

17. See *http://www.cites.org/*.

18. Philipp Weis, "New Rules Ease Specimen Shipments," *Science* 298 (22 November 2002): 1539.

19. For example, Mathis Wackernagel and William E. Rees, *Our Ecological Footprint: Reducing Human Impact on the Earth* (Gabriola Island, BC; Philadelphia: New Society Publishers, 1996); Paul Ehrlich and Anne Ehrlich, *Extinction: The Causes and Consequences of the Disappearance of Species*

(New York: Random House, 1981); Paul Ehrlich, Anne Ehrlich, and Gretchen Daily, *The Stork and the Plow: The Equity Answer to the Human Dilemma* (New York: Putnam, 1995).

20. E. F. Schumacher, *Small Is Beautiful: Economics as If People Mattered* (New York: Harper Perennial, 1989).

21. Denton Morrison, "Soft Tech/Hard Tech, Hi Tech/Low Tech: A Social Movement Analysis of Appropriate Technology," *Sociological Inquiry* 53, nos. 2–3 (Spring–Summer 1983): 220–251.

22. Laughlan T. Munro, "Technology Choice in Bhutan: Labour Shortage, Aid Dependence, and a Mountain Environment," *Mountain Research and Development* 9, no. 1 (1989): 15–23.

23. Ibid.

24. Bernhard Glaeser and Kevin D. Phillips-Howard, "A Technological Alternative for Energy Use in Rural Development: The Case of Southeast Nigeria," *Journal of Public and International Affairs* 5, no. 1 (Winter 1984): 16–30. The authors indicate other energy savings in food preservation (use of shade, and drying in sacks, under roof rafters, and in clay pots rather than refrigeration).

25. Justus Laichena and James Wafula, "Biogas Technology for Rural Households in Kenya," *OPEC Review* 21 (1997): 223–244.

26. Ibid.

27. Andrew Barnett, "The Diffusion of Energy Technology in the Rural Areas of Developing Countries: A Synthesis of Recent Experience," *World Development* 18, no. 4 (1990): 539–553. In the Philippines, RETAIN specialists promoted charcoal gasifiers as a substitute for diesel in irrigation pumps. Another project involved decentralized electricity supply in Argentina through windmills.

28. Ibid.

29. Donald Heyneman, "Dams and Disease," *Human Nature* 2, no. 2 (December 1979): 50–57. Heyneman writes: "One of the tragedies of modern technology is that we are incapable of predicting or controlling its ecological impact." He continues by saying that no matter who drives the innovation, it "sometimes seems to create two new problems for every one it solves."

30. Garrett Hardin, "The Tragedy of the Commons," *Science* 162 (1968): 1243–1248.

31. Melissa Leach and Robin Mearns, "The Lie of the Land," in Leach and Mearns, editors, *The Lie of the Land* (London: Villiers Publications), 12.

32. Donella Meadows, Dennis Meadows, Jorgen Randers, and William

Behrens, *Limits to Growth: A Report to the Club of Rome Project on the Predicament of Mankind* (New York: Universe Books, 1974).

33. See *http://fletcher.tufts.edu/multi/www/1994-decl.html.*

34. "INDIA: Access to Clean Water Basic Right: Study 15–04–2002," at *http://www.ahrchk.net/news/mainfile.php/ahrnews_200204/2531/.*